Next-Generation Greenhouses for Food Security

Edited by Redmond R. Shamshiri

Published in London, United Kingdom

IntechOpen

Supporting open minds since 2005

Next-Generation Greenhouses for Food Security
http://dx.doi.org/10.5772/intechopen.92515
Edited by Redmond R. Shamshiri

Contributors
Aliyu Idris Muhammad, Musa Abubakar Tadda, Abubakar Shitu, Umar Abdulbaki Danhassan, Muhammad Hilal Kabir, Attanda Muhammed Lawal, Erick K. Kiplangat Ronoh, Seyed Moin-E-Ddin Rezvani, Redmond R. Shamshiri, Hamid Zare Abyane, Mohsen Godarzi, Davood Momeni, Siva Kumar Balasundram, Ibrahim A. Hameed, David Leroux, Mark Lefsrud, Kelly R. Thorp, Siva K. Balasundram, Sanaz Shafian, M. Fatemieh, Muhammad Sultan, Benjamin Mahns, Saba Samiei, Arasu Sivagami, Bhaskarrao Yakkala, Michael Angelo Kandavalli, Sunday J. Ojolo, Gbeminiyi G. Sobamowo, Hadeed Ashraf, Takahiko Miyazaki

First published in London, United Kingdom, 2021 by IntechOpen
IntechOpen is the global imprint of INTECHOPEN LIMITED, registered in England and Wales, registration number: 11086078, 5 Princes Gate Court, London, SW7 2QJ, United Kingdom
Printed in Croatia

British Library Cataloguing-in-Publication Data
A catalogue record for this book is available from the British Library

Additional hard and PDF copies can be obtained from orders@intechopen.com

Next-Generation Greenhouses for Food Security
Edited by Redmond R. Shamshiri
p. cm.
Print ISBN 978-1-83968-075-5
Online ISBN 978-1-83968-076-2
eBook (PDF) ISBN 978-1-83968-080-9

For EU product safety concerns:
IN TECH d.o.o., Prolaz Marije Krucifikse Kozulić 3, 51000 Rijeka, Croatia, info@intechopen.com or visit our website at intechopen.com.

Meet the editor

Dr. Redmond R. Shamshiri holds a Ph.D. in agricultural automation with a focus on control systems and dynamics. He is a scientist at the Leibniz-Institut für Agrartechnik und Bioökonomie working toward digitization of agriculture for food security. His main research fields include simulation and modeling for closed-field plant production systems, LPWAN sensors, wireless control, and autonomous navigation. His work has appeared in over 100 publications, including peer-reviewed journal papers, book chapters, and conference proceedings. He is a member of the Adaptive AgroTech Consultancy Network and serves as a section editor and reviewer for various high-ranking journals in the field of smart farming.

Contents

Preface

Closed-field crop production systems by means of controlled environments and high-tech greenhouses have faced significant technical improvements in terms of structural design, resource management, decision support systems, simulation models, and automation-control systems. It is predicted that by 2050, more than 70% of the world's population will live in the cities. This scenario challenges researchers and greenhouse growers to incorporate digital technology and examine different innovative cultivation techniques in order to secure the supply chain of fresh fruits and vegetables. In some regions where land is scarce, conventional greenhouses are being replaced with vertical farms, roof-top greenhouses, plant factories, and modular agri-cube units for urban farming in order to respond to the food security of the increasing world population. The main objectives of these platforms are increasing productivity and reducing expenses in a sustainable manner. The next-generation greenhouses are expected to produce "twice as much food using half as many resources." To achieve this, engineering solutions and technological developments have been integrated with agricultural sciences to reduce carbon footprint and minimize the dependencies on energy, space, soil, water, and natural light.

For modern high-tech greenhouses to attain their objectives and keep the production competitive, specific attention needs to be paid to the technical aspects of automation and control systems, environmental control methods, structural design, energy management, and cultural practices. This presented book aims to expand and highlight these aspects from an academic perspective in separate chapters. In the first chapter, Shamshiri et al. demonstrate real-time monitoring and wireless automation instruments that are integrated with advanced algorithms and artificial intelligence for providing a flexible control on the greenhouse environment. The second chapter is dedicated to the fundamentals of microclimate control systems followed by an overview of the advances in the desiccant and evaporative cooling systems. According to Sultan et al., solar-operated desiccant-based evaporative cooling systems could be an alternate option for next-generation greenhouse air-conditioning. The third chapter demonstrates a real-world example, the Canadian Integrated Northern Greenhouse (CING), that provides an adaptive design solution for growing fresh food year-round for northern Canadians. According to Leroux and Lefsrud, using container farming, the combination of natural and supplemental light has the potential to reduce energy needs linked to lighting. Chapters 4 and 5 discuss radiation exchange in greenhouses, as well as the requirements and the challenges for soilless crop production. Various plant growth models and simulation analyses for dynamic assessment of crop-growth microenvironments prior to and during the actual cultivation are reviewed and summarized in Chapter 6. In Chapter 7, the kinetic modeling of combustion and gasification zones for embracing greenhouse effects through biomass gasification is demonstrated. Chapter 8 presents an affordable open-source prototype for automated irrigation and environmental monitoring that can be used for experimenting and validating different control algorithms.

To ensure food security and self-sustainability, the next-generation greenhouses should incorporate advances in controlled-environment agriculture, energy optimization models, crop models, artificial lighting, and benefits from the concepts of

IoT devices, web-based data sharing applications, smart sensors, and artificial intelligent control algorithms for automation of the whole system. Most of the solutions and strategies described in this book represent a small but valuable contribution of the greenhouse research community toward higher yield and quality, reducing production losses, improving the sustainability of closed-field cultivation, and preserving natural resources. However, in most cases, depending on the region and the crop to be cultivated, the cost of the high-tech greenhouses relative to the increase in yield and profitability is not clearly well known. A constant joint effort and collaboration between growers, policymakers, and researchers will strengthen such an effort to arrive at an accurate economic analysis and justification for the high start-up costs involved with the next-generation greenhouses.

Redmond R. Shamshiri
Leibniz Institute of Agricultural Engineering and Bio-economy,
Potsdam, Germany

Chapter 1

Greenhouse Automation Using Wireless Sensors and IoT Instruments Integrated with Artificial Intelligence

Redmond R. Shamshiri, Ibrahim A. Hameed, Kelly R. Thorp, Siva K. Balasundram, Sanaz Shafian, Mohammad Fatemieh, Muhammad Sultan, Benjamin Mahns and Saba Samiei

Abstract

Automation of greenhouse environment using simple timer-based actuators or by means of conventional control algorithms that require feedbacks from offline sensors for switching devices are not efficient solutions in large-scale modern greenhouses. Wireless instruments that are integrated with artificial intelligence (AI) algorithms and knowledge-based decision support systems have attracted growers' attention due to their implementation flexibility, contribution to energy reduction, and yield predictability. Sustainable production of fruits and vegetables under greenhouse environments with reduced energy inputs entails proper integration of the existing climate control systems with IoT automation in order to incorporate real-time data transfer from multiple sensors into AI algorithms and crop growth models using cloud-based streaming systems. This chapter provides an overview of such an automation workflow in greenhouse environments by means of distributed wireless nodes that are custom-designed based on the powerful dual-core 32-bit microcontroller with LoRa modulation at 868 MHz. Sample results from commercial and research greenhouse experiments with the IoT hardware and software have been provided to show connection stability, robustness, and reliability. The presented setup allows deployment of AI on embedded hardware units such as CPUs and GPUs, or on cloud-based streaming systems that collect precise measurements from multiple sensors in different locations inside greenhouse environments.

Keywords: LoRaWAN, Greenhouse, Datalogger, Internet of Things, AgroTech, Leaf wetness

1. Introduction

Control and automation of microclimate and fertigation inside greenhouses have contributed to improving the sustainability of closed-field environment agriculture by reducing water, fertilizer, and energy demand, while at the same time increasing yield and profit [1]. The trend of environmental monitoring in modern farming is towards shifting from offline systems to wireless and cloud-based data collection architecture [2]. Advances in sensing technology have made possible the best quality of greenhouse production with the capability of yield prediction. Digital technology such as the Internet-of-Things (IoT) offers parallel solutions for automation engineers, which can be customized specifically for greenhouse applications. Wireless sensors and IoT enabled devices are used for real-time monitoring and control of the greenhouse environment through a secure internet connections on any mobile devices [3]. With multiple sensors that transmit data to a central computer installed with knowledge-based automation software, growers can monitor all internal and external data and apply any required changes to the environment in real-time. For example, a fertigation control system that monitors certain aspects of the irrigation, such as flow rate, electrical conductivity (EC), and pH of the fertigation solution, as well as the external variables such as solar radiation and external climate conditions can take advantage of the collected data and incorporate them into models or artificial intelligence algorithms in a way that particular control commands, such as triggering specific pumps or switching other processes, are sent to alter the greenhouse environment. In this aspect, the flexibility of the monitoring system and the knowledge behind the control algorithms are the key factors for an effective automation system. **Figure 1** shows a general architecture of wireless communication for IoT monitoring and control of multiple greenhouses. The main justifications for the deployment of such infrastructure can be summarized as (i) to provide real-time monitoring of the changes and variations to ensure optimal growth environment and minimize the risk of equipment malfunction, (ii) to share data with cloud-based decision support systems, and (iii) to send instant responses to the wireless actuators for reducing input costs and increasing yield and quality.

Research and development for adopting wireless communication technology in monitoring and control of greenhouse environments began in the late 1990s and early 2000s. One of the earliest reports of WSN application in greenhouse environment monitoring can be found in the work of [4]. The compact size, reliability, and cost-effectiveness of WSN modules, as well as flexibility for developing custom applications beside easy installation, have made this technology gain importance and popularity for Closed-Field Environment Agriculture (CFEA). Various remote systems, both prototype and commercial, have been designed for investigating functionalities and limitations inside greenhouses.

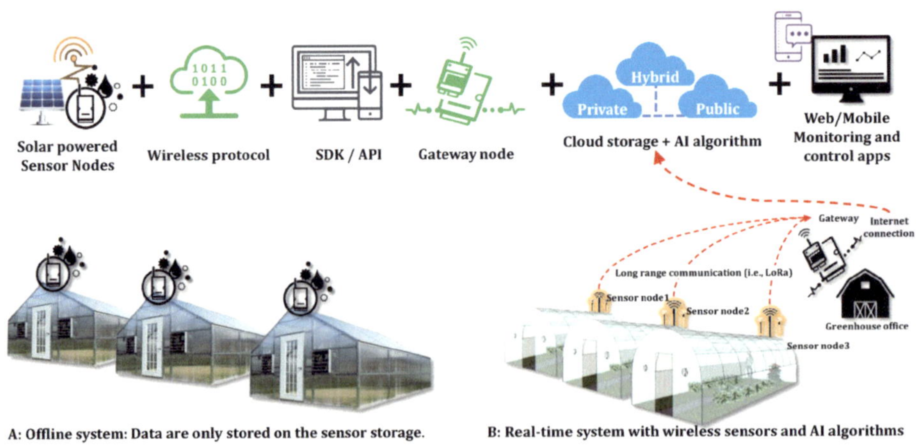

Figure 1.
Schematic diagram of wireless communication between greenhouse sensor nodes and cloud storage. Image by courtesy of Adaptive AgroTech.

An effective IoT-based solution should incorporate the use of wireless sensors and mobile applications for displaying, processing, and analyzing data from remote locations using cloud services which together provide new insights and recommendations for better decision-making. Evaluation of greenhouse environments prior to the actual cultivation is also of interest for many growers. IoT-based monitoring systems have been used for evaluating and adjusting microclimate parameters with LoRa sensors which are custom-designed to withstand hot and humid condition, allowing the system to continuously operate on solar-charged battery in remote areas where connections stability is of concern [2, 3]. An example of a modular LoRaWAN sensor node with external solar-charged battery and aviation connector cables with plug-and-sense capability is shown in **Figure 2**. These devices are customized specifically to operate in harsh agricultural condition and resist high humidity, solar radiation, insects, and bugs. The quality of network connectivity and stability in continuous data collection with 5 seconds intervals were tested in extreme conditions a proof of reliability for use in digital agriculture applications. A sample of air temperature dataset that was collected from a heat control chamber experiment using these devices are plotted in **Figure 3** to show the resolution and stability of the wireless transfer.

It should be noted that in most studies that are related to wireless monitoring of greenhouses, raw data are first collected via a wireless sensor network-based system and are processed afterward. A drawback of this approach is that because the collected data is not processed in real-time, they cannot immediately determine the temporal and spatial variations in the environmental parameters, as well as

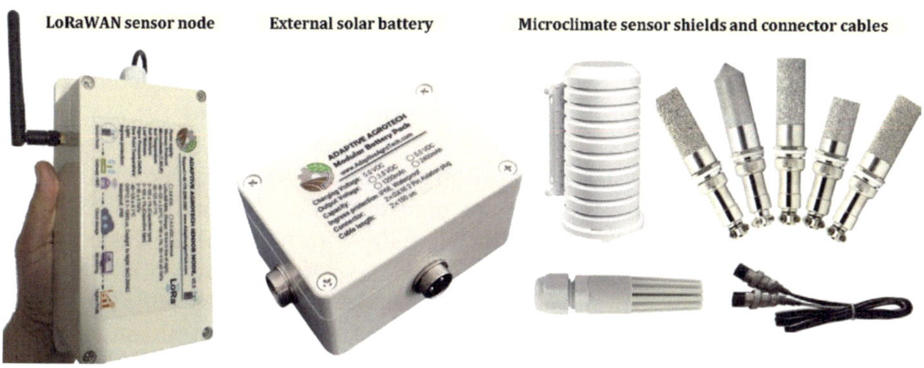

Figure 2.
A LoRaWAN wireless sensor node with an external solar-charged battery and different sensor shields used in real-time monitoring of greenhouse microclimate parameters. Images by courtesy of Adaptive AgroTech.

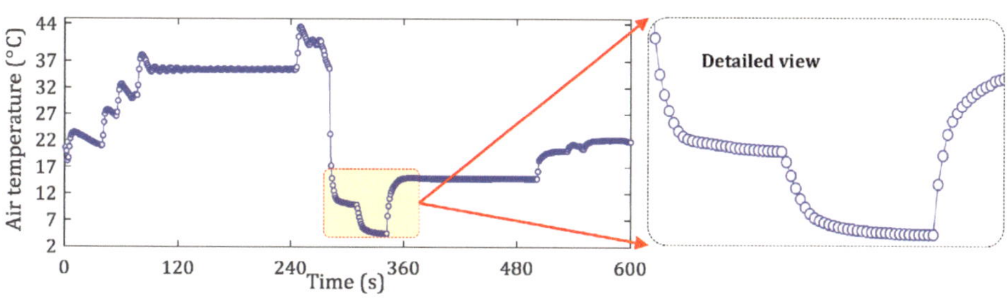

Figure 3.
An example of IoT monitoring of air temperature using ADP-AgroTech 868Mhz LoRa sensor located inside a metal heat control chamber that was isolated in a concrete basement for connectivity test. The gateway receiver was located inside another building, approximately 50 m away from the transmitter. Data were collected every 5 seconds.

their deviation from optimal conditions. In addition, the available commercial LoRaWAN and IoT sensors for use in agriculture and greenhouses environments are expensive and range between 1000 and 5000 USD at minimum order of 5 units. The presented chapter is an effort to respond to these problems, by presenting an overview of the components of an affordable multichannel wireless sensor node (WSN) with LoRa modulation at 868 MHz that can be interfaced with onboard computers such as Raspberry Pi for implementation of artificial intelligence (AI) algorithms in a way that they can perceive the greenhouse environment, make decisions, and take proper actions. These devices are custom-designed to withstand harsh greenhouse condition in order to provide real-time monitoring and control of crop growth variables such as microclimate parameters, light condition, soil temperature, soil moisture, and leaf wetness.

2. IoT sensing and data sharing in greenhouse production

Enabling commercial greenhouses with continuous sensing, communication between devices, and data sharing with the greenhouse management system is essential for disease prevention [5]. Some of the greenhouse diseases such as mildew fungi can cause significant loss of yield up to 50% [6]. For example, in hot and humid tropical climate conditions, extensive rainfall, fog, and high air temperature contribute to exacerbating the development of fungi in the leaves [7]. IoT-based sensor data fusion integrated with mathematical models provides growers with the opportunity to have a prediction of the situation and apply the right actions before an outbreak. The main elements of an IoT-based data acquisition and data sharing system with multiple sensor nodes and repeaters are shown in **Figure 4**. This framework provides growers with an evaluation of microclimate parameters with respect to different greenhouse designs and covering materials prior to the actual cultivation. The physical layer, software, and sensors layer in this scheme are linked wirelessly through standard communication protocols for transmitting data to a central base station for real-time or offline processing. This approach is required to exhibit precision accuracy, connection reliability within the sensing coverage, and low power consumption in order to be considered efficient for continuous monitoring of greenhouse in all growing seasons. Other than the specifications and characteristics of the sensors and communication algorithms that influence these functional properties, the physical internal and external condition of the greenhouse environment can also affect such a wireless monitoring framework.

2.1 Sensor probes

A sensor probe refers to any instrument or device that measures some physical or chemical characteristics of the environment and sends the results as an electrical

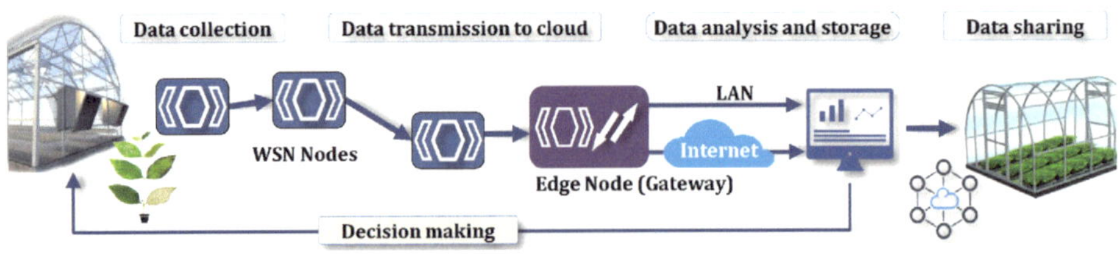

Figure 4.
Major components of an IoT-based data acquisition and monitoring system for greenhouse environment [2].

signal to be received and interpreted by the main automation computer for decision making and control purposes. For example, a pH sensor that continuously measures the pH of the irrigation water will trigger an alarm and maintain optimum pH level if it is too high or too low, eliminating the need for a grower to manually run pH tests and pH control. Some of the essential parameters to measure in a greenhouse environment include microclimate (i.e., air temperature, relative humidity, and vapor pressure deficit), soil moisture, soil temperature, and light level (or solar radiation) [6, 8–10]. Measurements from these sensors can indicate the presence of mildew disease or condensation. **Figure 5** shows some of the most widely used sensors in greenhouse production. For example, concerning the microclimate parameters, the high precision BlueDot BME280 + TSL2591 is a tiny integrated digital and cost-efficient sensor with great accuracy and range that provides a flexible solution. The BME280 is a combined digital humidity, pressure, and temperature sensor based on proven sensing principles. This sensor module is housed in an extremely compact metal-lid LGA package with a footprint of only 2.5 × 2.5 mm^2 with a height of 0.93 mm. Its small dimensions and its low power consumption allow the implementation in battery-driven sensor nodes inside greenhouses and can achieve high performance and accurate measurement. The BME280 also provides an extremely fast response time for fast context awareness applications and high overall accuracy over a wide temperature range. The pressure sensor is an absolute barometric pressure sensor with extremely high accuracy and resolution and drastically low noise. The integrated temperature sensor has been optimized for low noise and high resolution. Its output is used for temperature compensation of the pressure and humidity sensors and can also be used for estimation of the ambient temperature.

Most soil moisture sensors such as 10HS measures the dielectric constant of the soil using capacitance technology in order to find its volumetric water content (VWC), for scientific research and greenhouse applications. These sensors usually use 70 MHz frequency, which minimizes salinity and textural effects, providing high-resolution measurements that allow daily or hourly tracking of soil moisture content by sending analog voltage that is proportional to water content. These sensors have low sensitivity to salt and temperature, and are low power consumption. They can be connected directly or via interfaces to IoT boards for real-time

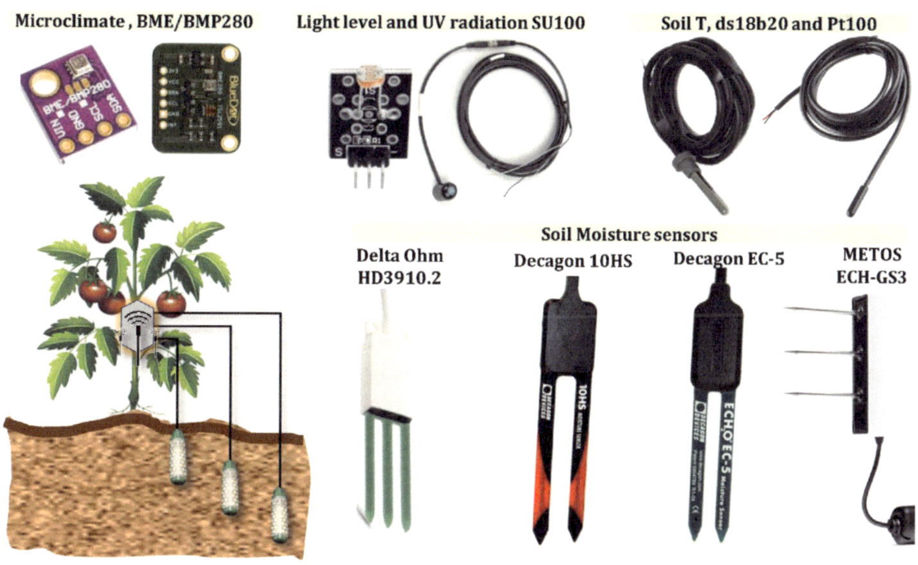

Figure 5.
Typical sensor probes used for measuring environmental variables in greenhouse crop production.

monitoring. Light level sensor, also known as Light Dependent Resistor (LDR) is an active sensor that is made of high accuracy, fast response, high resistance semiconductor which is sensitive to light. It decreases resistance with respect to receiving luminosity (light) on the component's sensitive surface. The resistance of a photo-resistor decreases with increasing incident light intensity (it exhibits photoconductivity). In the dark, a photo-resistor can have a resistance as high as several megaohms (MΩ), while in the light, a photo-resistor can have a resistance as low as a few hundred ohms. It should be noted that the raw output data from this sensor need to be calibrated for specific interpretation. The SQ-110 sensor, specifically calibrated for the detection of solar radiation, provides at its output a voltage proportional to the intensity of the light in the visible range of the spectrum, a key parameter in photosynthesis processes. The waterproof DS18b20 is a robust and corrosion-free sensor that can be used for measuring soil temperature. This sensor comes with different cable lengths of 1.8 and 3 m and provides 9-bit Celsius temperature measurements. The DS18B20 communicates over a 1-Wire bus that by definition requires only one data line (and ground) for communication with the connectivity board. Another temperature sensor, Pt – 1000, works based on the resistance that varies between approximately 920 Ω and 1200 Ω in the range considered useful in greenhouse applications (−20 ~ 50°C approximately), which results in too low variations of voltage at significant changes of temperature for the resolution of the analog-to-digital converter. Most soil moisture sensors are in fact analog sensors (non-rust capacitive hygrometer) that determine volumetric water content (VWC) by measuring the dielectric constant of the media using capacitance/frequency domain technology. An example is the analog sensor from Sun3Drucker that can be inserted directly into the soil to send moisture feedback data in real-time using capacitive sensing. A cable length of 1.5 m has been tested and found to be noise-free for these sensors. The soil moisture sensor probe is corrosion-free (no electrolysis on the electrodes) since it is using capacitive measuring method, and therefore is free of electrolysis on the electrodes. Another soil moisture sensor, the ECHO EC-5, determines volumetric water content (VWC) by measuring the dielectric constant of the media using capacitance/frequency domain technology. The EC-5 probe 70 MHz frequency minimizes salinity and textural effects, making this sensor accurate in almost any soil or soilless media. Factory calibrations are included for mineral soils, potting soils, Rockwool, and perlite.

Other than the mentioned sensors, some specific applications in greenhouse production and research may require a custom-design sensor probe. For example, in a greenhouse with misting or fogging systems, it is necessary to determine the solution droplet deposition on the plants. Determining leaf wetness as a reference measurement to avoid condensation inside greenhouse environments in certain hours is also of interest. In large-scale commercial greenhouse production measuring leaf surface wetness to determine the performance of spraying is required for chemical depletion. For this purpose, the ADP-AgroTech leaf wetness sensor model ADP-LWS2020 shown in **Figure 6** has been designed with different shapes to mimic the actual leaf shape, and to convert the moisture on the leaf surface into an analog signal using capacitance change. This sensor has been optimized to eliminate noise and generate high-resolution output under extreme greenhouse conditions. The performance of this sensor has been tested under high temperature and humidity in different tropical lowlands of Malaysia, and has been found to be stable and resistant under direct solar radiation. It can be seen from **Figure 6** that the surface of this sensor is composed of several rows of dielectric constant capacitor that has equal spacing and are connected to an electronic interface board for producing an analog signal. The ADP-LWS2020 can mimic the wetness state of a real leaf and detects the presence of surface moisture and calculates the duration of wetness.

Figure 6.
ADP-AgroTech leaf wetness sensor (model ADP-LWS2020) with different leaf shapes based on capacitive method for determining leaf surface moisture and greenhouse condensation. Images by courtesy of Adaptive AgroTech

The voltage at its output is inversely proportional to the humidity condensed on the sensor, and can be read at the analog input of Adaptive AgroTech connectivity boards. It can be used for greenhouse studies and control systems and for scheduling irrigation. It also allows researchers to protect plants by giving early warnings about fungus and insect attacks.

2.2 IoT connectivity boards and modular accessories

Different multi-channel connectivity boards with WiFi and LoRa antenna that benefits from a modular design to be easily interfaced with sensor probes are shown in **Figure 7**. These boards are custom-designed to make possible adding new sensing capabilities to the existing wireless networks with minimum effort. In the same way, defective sensor probes may be easily replaced in order to ensure the lowest maintenance cost of the sensor network. The connectivity boards that are shown in **Figure 7** include all the electronics and sockets necessary to connect the most typical sensors in wireless monitoring of greenhouse environment, including BME280 (air temperature, humidity, and atmospheric pressure), DS18B20 (soil temperature), LDR Photoresistor (light sensor), SX239 (soil moisture), and NEO-7 GNSS modules. The custom-designed version of these boards include the necessary components for more specific research applications, such as Pt-1000, ADP-AgroTech leaf wetness (shown in **Figure 6**), weather station (pluviometer, anemometer, and vane), Luminosity sensor (TSL2561), and distance sensor (TFmini from Benewake). For more robust and fast processing, the connectivity boards in **Figure 7** benefit from the powerful ESP32 and Atmega328P microcontrollers that are integrated with customized codes for high efficiency and ultra-low power consumption (deep-sleep mode). The wireless communication between

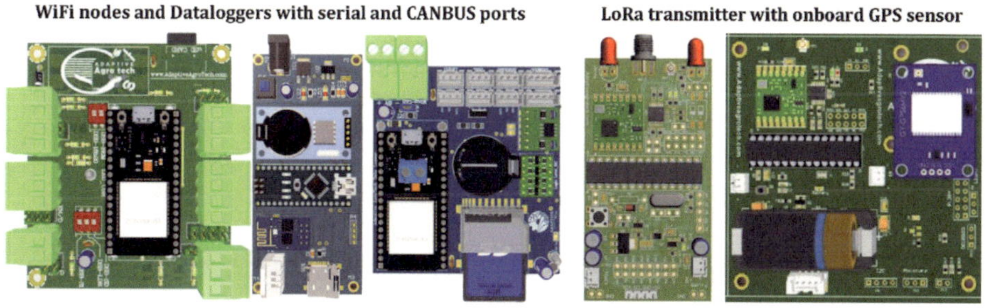

Figure 7.
Sample prototype of WiFi and LoRa connectivity boards with onboard storage for real-time monitoring and IoT control of greenhouse based on ESP32 and Atmega328P microcontroller. Images by courtesy of Adaptive AgroTech.

these transmitter boards and receiver (gateway) is realized through Lora technology (433 MHz (Asia), 868 MHz (Europe) and 915 MHz (Australia and North America)) which covers 2 ~ 10 km distance in rural areas and is extendable to 100 km with repeaters. By default these boards have been programmed to read and record measurements every 10 seconds which can be adjusted according to the growers' needs. Data are stored on an onboard mini SD card or are transferred to an open-source secure cloud database via WiFi connection. Data can be viewed online at http://iot.adaptiveagrotech.com/ or by installing Adaptive AgroTech smartphone app.

The wireless sensor and controller boards as well as other electronic components and modules are housed in sealed waterproof ABS enclosures that are rated as IP-66 in order to withstand harsh environmental conditions such as sunlight, dust, moisture, high humidity, insects, and sudden changes in temperature. **Figure 8** shows a hybrid data acquisition system with modular components including the main connectivity board, sensor probes, connector cables, external solar-charged battery, and a solar panel. These components are interchangeable and can be connected to the mainboard using standard aviation plug GX16 male and female metal sockets. For greenhouse application, it is very important that all enclosures are high-quality ABS flame retardant material, corrosion resistance, anti-UV and anti-aging, antistatic, good sealing performance, long life, and suitable for all types of environments. The performance of these components and the metal sockets has been tested for over 12 months in different open-field and closed-field agriculture production. The external battery shown in **Figure 8** is 5.0 V, 2400mAh that can be continuously charged with a 5 V, 500mAh solar panel, and can last over two years without any maintenance at 60 readings per hour when the mainboard is operating in deep-sleep mode. The voltage of the battery can be adjusted and increased to 7.7 V or reduced to 3.8 V for other applications. It is recommended that those sensor probes that are not intended to be used during the data collection should not be connected to the boards. Since several sensors share the same power line, a sensor that is not going to be used and still connected to the board will entail an additional consumption, resulting in a shorter life of the battery. **Figure 8** also shows that the microclimate sensor has been placed in a protective shell to withstand direct sunlight and moisture, and to stabilize the air temperature and relative humidity for more accurate measurement and preventing errors. Other types of shells for microclimate sensors can be used for greenhouse depending on the application. These shells are waterproof and will keep water from seeping into the body of the sensor and damaging it, while at the same time airflow can pass through.

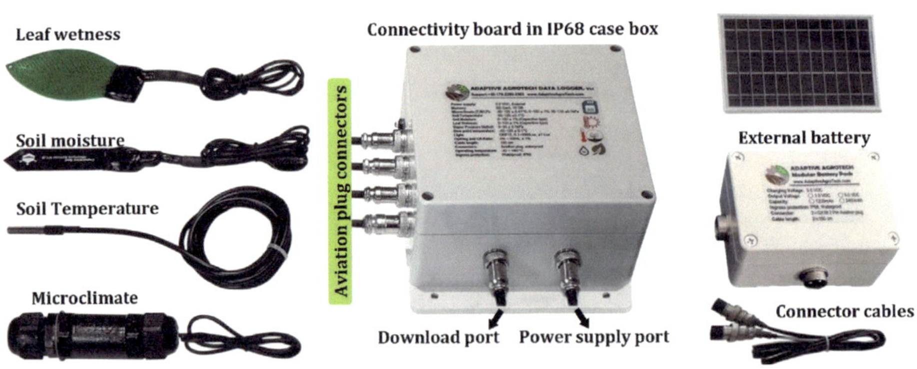

Figure 8.
A hybrid data acquisition system with modular solar charged external battery, plug-and-sense probes, 32GB onboard, and multiple communication interfaces for data transfer including serial port, WiFi, and LoRa 868Mhz. Image by courtesy of Adaptive AgroTech.

2.3 Wireless communication and IoT-based monitoring and control

The trend in the monitoring of environmental parameters inside modern greenhouses is towards shifting from offline systems to wireless and cloud-based data collection architecture. Various remote systems, either by means of prototype or commercial, have been used for improving the performance of greenhouse monitoring. Some of the most recent examples include web-based, cloud-based, and IoT data collection, monitoring and control system [2, 3], wireless sensor networks [5, 8], field server-based monitoring [11], field router systems [12], and distributed data acquisition with a local controller and management [9]. A comprehensive comparison between the existing remote monitoring system in agricultural research is available in the work of [13]. It should be mentioned that the core part of any IoT sensing and control system is the wireless communication between the devices and the internet. A summary of the wireless communication that can be used in the greenhouse industry is presented in **Table 1** to provide a quick comparison between their frequency bands, sensitivity, and coverage range. Network health analyzer software can also be used to check data transfer reliability.

IoT-based monitoring solutions [3] reduce data collection errors in greenhouse environments, while at the same time increase the flexibility of the remote control of devices. Real-time data generated from this process enables growers to have a continuous evaluation of the crop growth environment through dynamic assessment. The traditional techniques frequently suffer from great labor intensity, low spatiotemporal resolution, a lack of mechanization and organization and also needing much time in the growing of plants and observing the environmental aspects of the greenhouse. To address these problems, an IoT controller board and a modular wireless Datalogger system shown in **Figure 9** were custom-designed to provide communication between sensor nodes, end-users, and greenhouse actuators. The controller has an onboard Raspberry Pi computer and two microcontrollers and is capable of receiving command signals using WiFi connection to run an 8-channel relay board, and two motor drivers. The control signals can be either generated by the greenhouse crop models algorithms that are coded into the onboard computer, or by the cloud-based streaming systems. At the same time, environmental sensors can collect measurements, store data on a SD card, and transmit data directly to a web-server, or via wireless communication to a gateway using LoRa 868Mhz frequency. This platform allows real-time monitoring of the data on Adaptive AgroTech private secure cloud system which is accessible at iot.adaptiveagrotech.com or by installing the mobile application. A detailed description of this platform is available in [3]. Some of the specific application of the modular and flexible IoT automation system shown in **Figure 9** can be summarized as: multi-purpose application for real-time monitoring in closed-field and open-field agriculture, measuring optimality degree and comfort ratio of greenhouse environments, as well as yield prediction of tomato using Simulink blocks and embedded crop growth models, prevention of plant diseases based on predictive models, multiple voltage lines for DC actuators, 8-channel relay controller, two stepper and DC motor drivers, open-source programming, LoRaWAN connectivity with built-in light sensor, GPS, and microclimate sensor, and waterproof IP66 enclosure with external battery module and charging circuits.

The architecture of the data transmission from sensor nodes to cloud-storage and from web-server to the controller is shown in **Figure 10**. A total of four layers, including the farm layer (with sensor nodes), the backend layer, the wrapper later, and the frontend later are integrated in a way that end users can access data from their phone or desktop applications for real-time monitoring of the sensor measurements. In this scheme, each request sender is treated as the client, and the response provider as the server. The farm later has the role of (i) provider, in which wireless sensor nodes in

Radio	Protocol	Frequency bands	Transmission power	Sensitivity	Range*	Certification
XBee-PRO 802.15.4 EU	802.15.4	2.4 GHz	10 dBm	-100 dBm	750 m	CE
XBee-PRO 802.15.4	802.15.4	2.4 GHz	18 dBm	-100 dBm	1600 m	FCC, IC, ANATEL, RCM
XBee 868LP	RF	868 MHz	14 dBm	−106 dBm	8.4 km	CE
XBee 900HP US	RF	900 MHz	24 dBm	−110 dBm	15.5 km	FCC, IC
XBee 900HP BR	RF	900 MHz	24 dBm	−110 dBm	15.5 km	ANATEL
XBee 900HP AU	RF	900 MHz	24 dBm	−110 dBm	15.5 km	RCM
WiFi	WiFi (HTTP(S), FTP, TCP, UDP)	2.4 GHz	17 dBm	−94 dBm	500 m	CE, FCC, IC, ANATEL, RCM
4G EU/BR	4G/3G/2G (HTTP, FTP, TCP, UDP) GPS	800, 850, 900, 1800, 2100, 600 MHz	4G: class 3 (0.2 W, 23 dBm)	4G: −102 dBm	- km - Typical base station range	CE, ANATEL
4G US	4G/3G/2G (HTTP, FTP, TCP, UDP) GPS	700, 850, 1700, 1900 MHz	4G: class 3 (0.2 W, 23 dBm)	4G: −103 dBm	- km - Typical base station range	FCC, IC, PTCRB, AT&T
4G AU	4G (HTTP, FTP, TCP, UDP)	700, 1800, 2600 MHz	4G: class 3 (0.2 W, 23 dBm)	4G: −102 dBm	- km - Typical base station range	RCM
Sigfox EU	Sigfox	868 MHz	16 dBm	−126 dBm	km - Typical base station range	CE
Sigfox US	Sigfox	900 MHz	24 dBm	−127 dBm	km - Typical base station range	FCC, IC
Sigfox AU / APAC / LATAM	Sigfox	900 MHz	24 dBm	−127 dBm	km - Typical base station range	—
LoRaWAN EU	LoRaWAN	868 MHz	14 dBm	−136 dBm	> 15 km	CE
LoRaWAN US	LoRaWAN	902-928 MHz	18.5 dBm	−136 dBm	> 15 km	FCC, IC

Radio	Protocol	Frequency bands	Transmission power	Sensitivity	Range*	Certification
LoRaWAN AU	LoRaWAN	915-928 MHz	18.5 dBm	−136 dBm	> 15 km	—
LoRaWAN IN	LoRaWAN	865-867 MHz	18.5 dBm	−136 dBm	> 15 km	—
LoRaWAN ASIA-PAC/ LATAM	LoRaWAN	923 MHz	18.5 dBm	−136 dBm	> 15 km	—

Line of sight and Fresnel zone clearance with 5dBi dipole antenna.

Table 1.
Standard wireless communications used in agricultural applications.

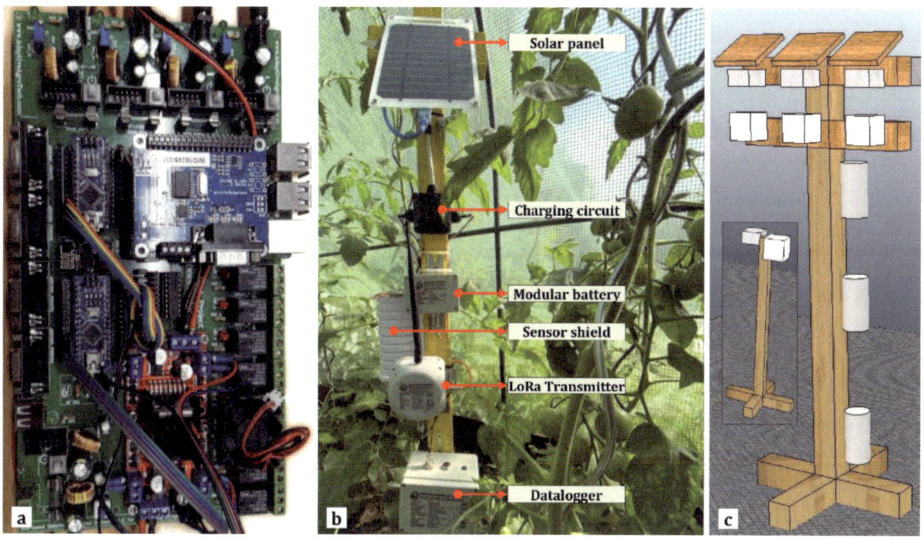

Figure 9.
A prototype IoT automation system used in real-time monitoring and control of greenhouse environments, (a) a controller board with two WiFi modules, onboard computer, 8-channel relays, and 2 stepper motor drivers, (b) a wireless LoRa sensor and Datalogger with modular components, (c) a custom-built sensor platform.

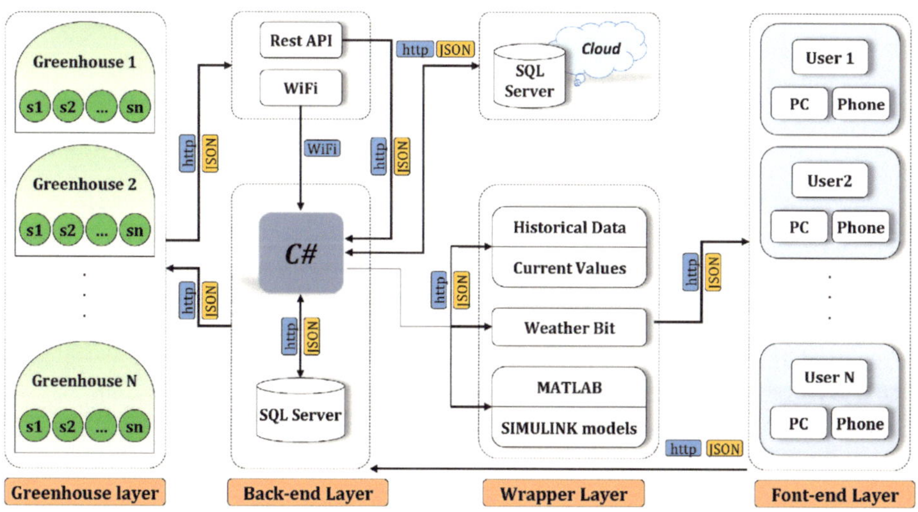

Figure 10.
A general architecture of a WSN based monitoring of greenhouse environment.

the farm collect data and transmit to a gateway device that has access to the internet using WiFi protocol, or convert the data packet to JSON format before sending the data to the backend layer using HTTPS protocol, (ii) client, in which each wireless node sends requests to the backend and receive responses in JSON format via HTTP protocol. The backend layer consists of a middle layer between the backend server and the farm layer. A middle device or server in the backend layer that uses WiFi and REST API providers first receive data from the farm layer and then transfer the packets to the backend layer. The programming language used in the backend server is C# and the database is an SQL server. Received data are pre-processed, analyzed, and then categorized using queries, crop models, and AI algorithms, and are then saved in the database using controllers that have been implemented in the C# frameworks. The queries can run on the database to receive responses in the defined format. The communication between the backend server and SQL server is established using HTTP protocol and JSON format. The wrapper layer includes the cloud storage in which processed data from the backend are sent to IoT serve and are saved. This gives the user the advantage of having a secure backup of the collected data. The provider receives data from the backend layer and for further real-time assessment of the field condition. The input of this layer is the transferred data which are collected every 5 or 10 minutes by the field layer (sensor nodes in the farm). The frontend layer, also called the presentation layer, provides data visualization by means of real-time plots, control buttons, and indicators on, mobile apps, webpages, or other platforms. The frontend layer can have access to the collected data via the backend layer using send-and-request protocols, and receive responses from the wrapper layer. The presented wireless IoT framework was tested in various greenhouse environments and showed that while the sensor measurement was 100%, the network mean packet reliability was between 95 and 100% due to the packet losses. This failure can be related to the high-density plants canopy which can significantly reduce the signal strength of the sensor nodes. Graphical results of experimenting with the wireless sensor and IoT controllers are provided in **Figures 11** and **12**.

2.4 Case study: verifying performance of a crop model with a WiFi sensor node

In order to improve greenhouse yield and profits, collected data from multiple wireless sensors that are deployed in different parts of the greenhouse should be used with knowledge-based software, and crop growth models. These models are often sensitive to boundary inputs and may cause inaccurate simulation results. The objective of this case study was to use a WiFi sensor node for collecting air temperature and light data in order to evaluate parameter robustness of the reduced state-variable TOMGRO model [14] for yield estimation of tomato in a random greenhouse. The hypothesis was to test whether the model parameters are robust enough to translate an adverse greenhouse environment (with air temperature so high to prevent any crop growth development) to realistic biomass and yield. For this purpose, TOMGRO was first implemented in Matlab Simulink in order to create a flexible platform for easier interfacing with the inputs and outputs. The final Simulink block was validated with the Lakecity datasets of [14]. To produce boundary data, an experiment was carried out in an empty glass-panels covered greenhouse under tropical lowlands climate conditions by turning off all ventilation and cooling systems for creating an adverse microclimate scenario with zero yield expectation. The glasshouse was located at the campus of Malaysian Agricultural Research and Development Institute (Latitude: 2°59′24.7", Longitude: 101°41′56.1"). Hourly measurements of air temperature and solar radiation were continuously collected for 254 days using a WiFi sensor node similar to the one shown in **Figure 13**. Plots of raw air temperature and solar radiation data from the glasshouse experiment are also shown in **Figure 13** followed by a detailed outlook of

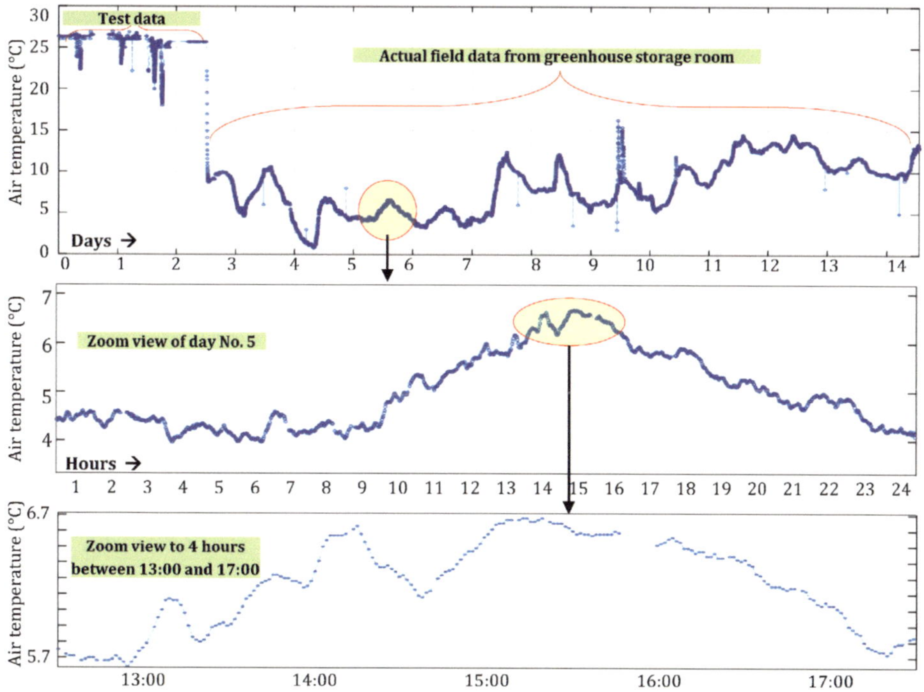

Figure 11.
Sample of air temperature data collected every 60 seconds using Adaptive AgroTech LoRaWAN sensor located inside a greenhouse storage room.

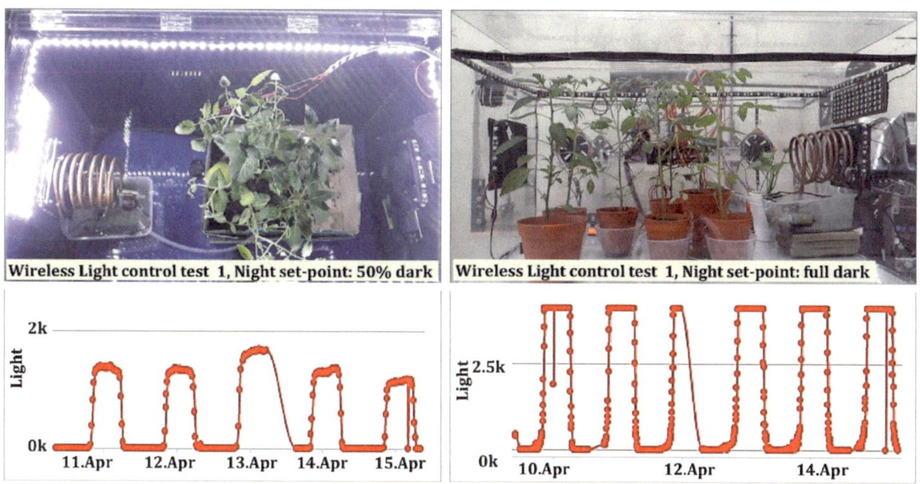

Figure 12.
Lab scale implementation of IoT monitoring and control of light level using LoRa 868 Mhz transceivers.

air temperature plots that shows daily maximum, averaged, and minimum values. It can be seen that air temperature readings inside the glasshouse have reached to 68°C in some specific hours, which not only prevents tomato growth, but can also have serious negative impacts on the solar-charged battery of the WiFi sensor node. In addition, daily averaged air temperature values between hours of 12:00 and 18:00 are in the range of 30 to 50°C. The average, minimum and maximum values during the entire experiment were equal to 34.5, 22.5, and 68.3°C, corresponding to a simulated growth response of zero between hours of 12:00 and 18:00. Results of simulation with

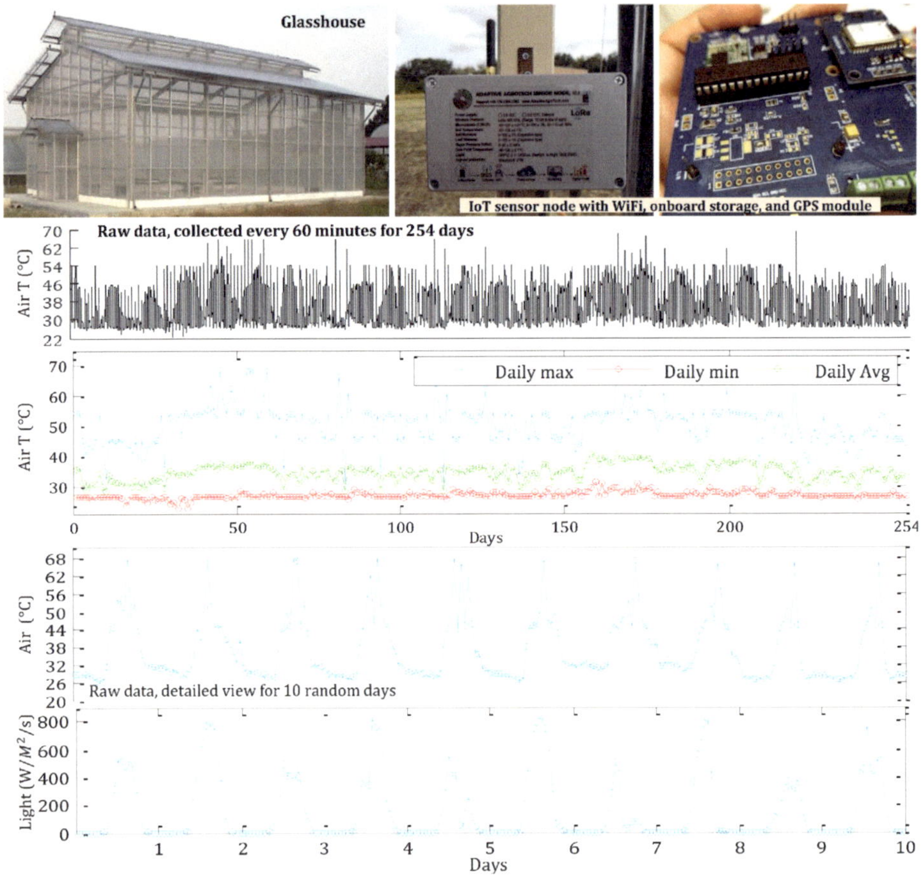

Figure 13.
Performance of IoT monitoring with WiFi sensor node in an empty glasshouse without climate control for testing the performance of the connectivity board and battery modules under adversely hot and humid environment.

TOMGRO model showed that the estimated total above-ground dry weight (W_T), total fruit dry weight (W_F), and mature fruit dry weight (W_M) were equal to 0.576, 0.085 and 0.072 kg/m^2 respectively. This study tested the claimed conclusion of [14] that states "it is possible to use the same reduced model with parameters estimated at one location to simulate leaf area and above-ground weight of tomato growing in greenhouse conditions in other locations" using boundary data that were continuously collected by a WiFi sensor node: Based on the consistency of the low estimated fruit yield with the simulated growth responses, the hypothesis that the simplified TOMGRO model with its initial parameters is not capable of estimating tomato yield for a random greenhouse in a different geographical location was rejected. It can be concluded that long-term historical data collected by IoT sensor nodes can be used to improve the performance of crop models, as well as offering new insights to add artificial intelligence algorithms to the automation system.

3. Artificial intelligence in greenhouse automation

Automation and control of greenhouse environments have to deal with various uncertainties and disturbances that cannot be entirely modeled by mathematical equations [1, 7, 15–17]. Adding artificial intelligence to greenhouse automation means that the AI algorithm must coexist with all other pieces of the automation system fluidly, including multiple sensors, physical systems that control devices and actuators. The main justification for designing AI-based automation systems in greenhouse environments such as those that operate with fuzzy logic or neural network algorithms is to shift toward a robust, predictive, and adaptive control command strategies that reduce production costs and improve yield. Artificial intelligence is a computer system that is programmed to present intelligent behavior by perceiving the environment, making decisions, and taking action. AI can contribute to sustainable greenhouse production in different ways such as reducing the electrical energy consumption of the climate control systems, or reducing water and chemical demands for fertigation system. For example, AI algorithms can be implemented for updating microclimate set-points (also known as reference values) depending on a specific crop, growth stages, light conditions, and external variables. These set-points are conventionally created manually by expert growers, or by means of knowledge-based decision support systems such as adaptive management [18] or dynamic assessment [19]. Set-points [6] are the core inputs of the microclimate control system and therefore must be calculated precisely, otherwise production failure and crop loss can occur in a few hours. Results of an experiment with three different tropical greenhouses in the lowlands of Malaysia that are shown in **Figure 14** reveal that without proper climate control algorithms, air temperature can reach 37°C or 60°C depending on the structural design and external condition [20, 21].

To overcome these challenges, a conventional greenhouse climate controller that triggers ventilation, misting, or spraying in order to reduce air temperature was developed and tested (**Figure 15**). During cold seasons, the controller was interfaced with time-based or sensor-based actuators for triggering of the heating system. These approached however are not efficient for high-tech large-scale greenhouses. An integrated climate control system should not only benefit from the wireless and IoT automation technology, but also from the innovative cooling and heating methods that operates based on AI algorithms. In this scheme, collected data from multiple wireless sensors that are deployed in different parts of the greenhouse are used to train machine learning algorithms that have been designed based on knowledge-based systems and mathematical crop growth models. The output commands and decision messages from this process are then used to control

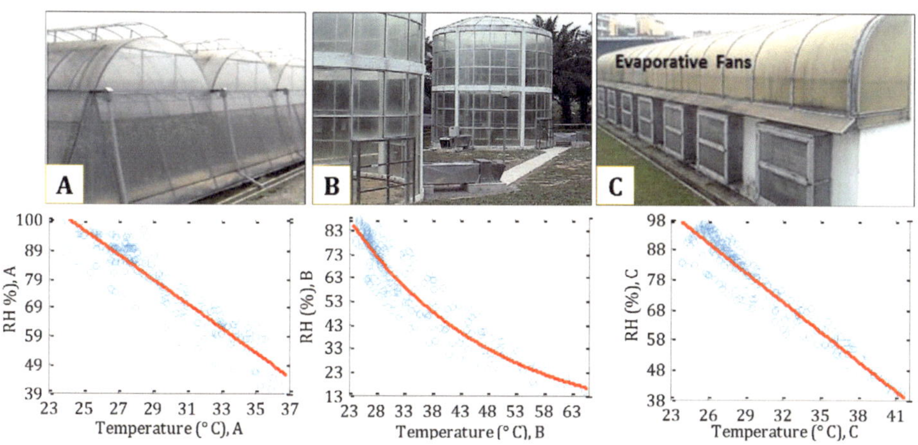

Figure 14.
Wireless monitoring of microclimate inside three tropical greenhouses with different covering materials (A: net-screen, B: polyethylene film, C: Polycarbonate panels) without proper climate control algorithms showing that air temperature are significantly far from optimal set-points and can exceed 37 °C or 60 °C depending on the structural design and outside condition.

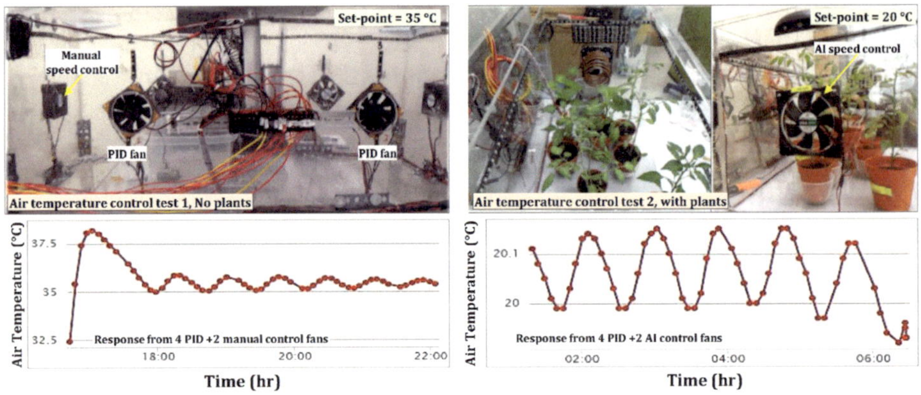

Figure 15.
Lab-scale implementation of IoT monitoring and control of air temperature, (left): 4 PID plus 2 manual-controlled fans, and (right): 4 PID plus 2 AI-controlled fans.

specific elements within the crop growth microenvironment while at the same time they are optimized to reduce energy, chemicals, and water demands. An example of this approach was implemented on a lab-scale greenhouse shown in **Figure 15** by performing PID speed control on 4 ventilation fans, and manual speed control on the other two fans (referred to as AI fans) for obtaining training dataset. The presented platform allowed experimenting with various methods, including fuzzy-logic self-tuning PID controller and machine learning to adjust the speed of the two AI fans. The air temperature responses were then monitored in real-time using WiFi sensor nodes and are shown by the two plots in **Figure 15**. A simple fuzzy logic control algorithm was also implemented on a research tropical greenhouse shown in **Figure 16** to demonstrate the difference between air temperature response in a timer-based control and intelligent control. A summary of the fuzzy logic rules is presented in **Tables 2** and **3**.

Another example of AI application in greenhouses is the prediction of micro-climate parameters as demonstrated in **Figure 17**. This prediction can be used for advanced microclimate control systems such as adaptive or predictive control, energy demand calculation, or for applications such as disease prevention, decision support systems, and cost–benefit analysis. It should be noted that building a successful AI algorithm for this purpose requires navigating the entire AI workflow and focusing on more than just one training data set and model. In this example, several datasets of the past 10 days from different tropical greenhouses were used to predict the 11th day data. Extensive simulations with different numbers of days were used to find out

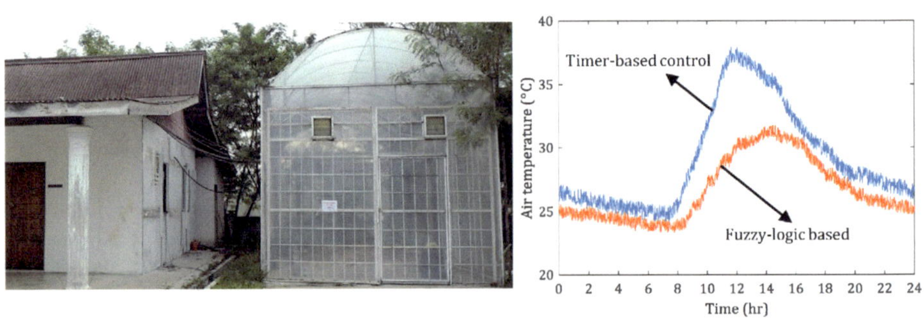

Figure 16.
A comparison between timer-based and fuzzy-logic based control of air temperature in an experimental greenhouse under lowland climate conditions of Malaysia.

State	$T_{setpoint}$	AirT$_{inside}$ $T_{(t)}$	AirT$_{inside}$ $T_{(t+1)}$	e_t $(T_s - T_{in\,(t)})$	e_{t+1} $(T_s - T_{in\,(t+1)})$	Δe	Output	
1	25	**30**	**30**	−5	−5	**0**	e = Negative, Δe = Zero	**Cool Normal**
2	25	30	29	−5	−4	+1	e = Negative, Δe = Positive	Stop Cool
3	25	30	31	−5	−6	−1	e = Negative, Δe = Negative	Cool Big
4	25	25	25	0	0	0	e = Zero, Δe = Zero	Ideal
5	25	25	24	0	+1	+1	e = Zero, Δe = Positive	Warm Normal
6	25	25	26	0	−1	−1	e = Zero, Δe = Negative	Cool Normal
7	25	20	20	+5	+5	0	e = Positive, Δe = Zero	Warm Big
8	25	20	19	+5	+6	+1	e = Positive, Δe = Positive	Warm Very Big
9	25	20	21	+5	+4	−1	e = Positive, Δe = Negative	Warm Normal

Table 2.
Example of a simple fuzzy logic control algorithm implemented on a research tropical greenhouse,: $T_s = T_{setpoint}$, $T_{in\,(t)} = T_{inside}$ *at Time (t)*, $T_{in\,(t+1)} = T_{inside}$ *at Time (t + 1)*, e_t *= error at Time (t)* $= T_s - T_{in\,(t)}$, e_{t+1} *= error at Time (t + 1)* $= T_s - T_{in\,(t+1)}$, *Δe = change of error =* $e_{(t+1)} - e_{(t)}$.

errorΔerror	Negative Big	Negative	Zero	Positive	Positive Big
Negative Big	Cool Big	Cool Big	Cool	STOP Cooling	Warm Normal
Negative	Cool Big	Cool	Cool	STOP Cooling	Warm Normal
Zero	Cool	Cool	Current condition	STOP Cooling	Warm Normal
Positive	STOP Cooling	STOP Cooling	Warm Normal	Warm Normal	Warm Big
Positive Big	STOP Cooling	Warm Normal	Warm Normal	Warm Big	Warm Big

Table 3.
The fuzzy logic rule table.

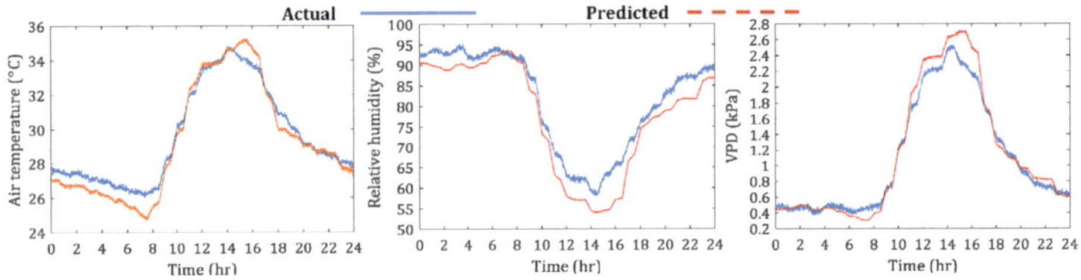

Figure 17.
Preliminary results of predicting the next 24 hours of air temperature, relative humidity, and vapor pressure deficit in a naturally ventilated tropical greenhouse using feed-forward neural network and a 10-days dataset.

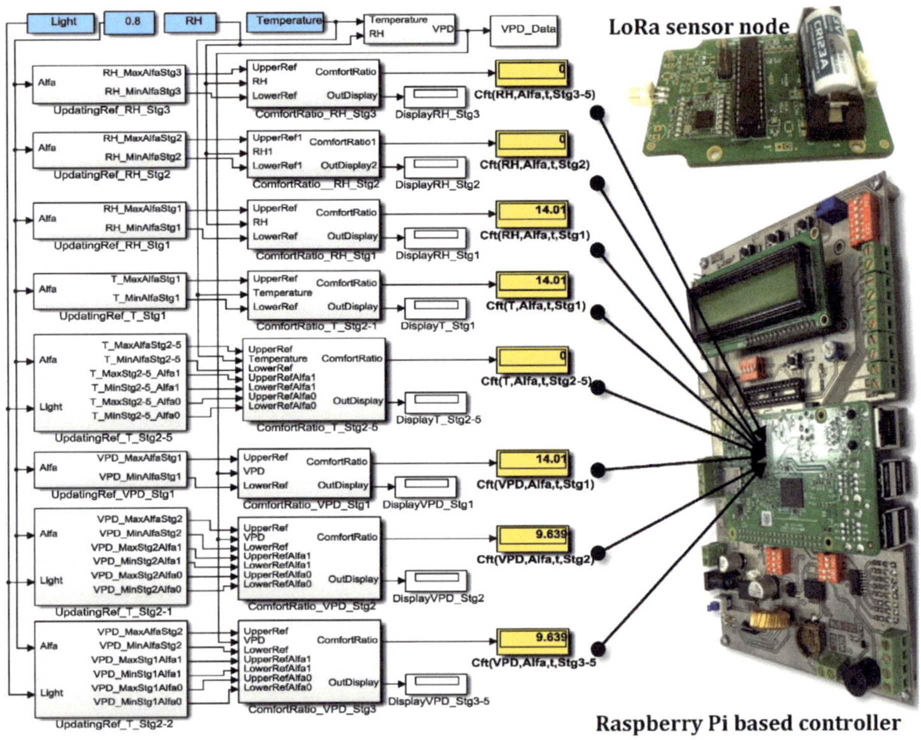

Figure 18.
IoT realization of the adaptive management framework using Simulink blocks, LoRa 868Mhz sensor node, and raspberry Pi embedded board for evaluation and adjusting greenhouse microclimate.

that a dataset that includes at least the past 10 days' measurement is required for the AI algorithm to effectively predict the microclimate of the next day. In this example, data preparation was more than having a lot of data or even pre-processing all of the data to be consistent. This process involved adding human insight to the selection of the training data, as well as considering augmenting data sets with synthetic data and more samples, and providing clean labeled data. In this regard, choosing the right AI algorithm, such as machine learning, deep learning, or a combination, and identifying the optimal set of parameters will lead to the most robust and accurate prediction model. As mentioned before, simulation techniques are extensively used to verify the performance of AI algorithms in every situation and scenario, such as different climate conditions, greenhouse structural design, covering materials, the crop that is being cultivated, and the growth stage. An example of simulation is the adaptive management framework [18] that allows growers to verify edge cases and test and run hundreds of scenarios that would otherwise be too time-and-cost intensive. In the example of microclimate prediction shown in **Figure 17**, the outputs of the AI which are microclimate data of the next 24 hours are used with predictive and adaptive control algorithms [18], therefore simulation enables validating the control process before deploying the codes on the actual hardware.

The final AI-based automation algorithms can be deployed as computer codes or Simulink blocks on cloud-based streaming systems, or on a local onboard computer similar to the one shown in **Figure 18**, which demonstrates IoT sensor fusion in combination with a comfort ratio model [2, 19, 20, 22] for real-time dynamic assessment of microclimate parameters in commercial scale greenhouse production of tomato. This method is based on the integration of wireless communication, distributed data analyzing and a web-based data monitoring dashboard that is used for data collection, processing, and monitoring. The wireless sensor node has shown a high spatiotemporal resolution with excellent stability in data transfer at 10 readings per minute within 1 km distance from the LoRaWAN gateway. The presented boards in **Figure 18** has been used as a proof-of-concept and showed the opportunity to use these new tools for model-based investigation of the spatial and temporal variations in the air temperature, relative humidity and, VPD inside greenhouse crop production [3]. The implication is to provide growers with digital tools that can assist in knowledge-based decision making for minimizing energy cost and yield loss due to low fruit quality. Moreover, the IoT automation system and cloud data processing contribute as a real-time online assessment tool to investigate effects of structure design, covering materials, cooling techniques, and growing seasons on the optimality and comfortability of microclimate parameters and their correlation with yields.

4. Conclusion

This chapter provided an overview of the application of IoT sensors and controllers that can be integrated with crop models and artificial intelligence algorithms for sustainable greenhouse production. Several affordable yet robust wireless sensor nodes developed by Adaptive AgroTech that benefit from WiFi and LoRa communication were presented with sample results from lab-scale and commercial-scale greenhouses. The introduced wireless transceivers were shown to be flexible and modular, which makes possible easy installation anywhere in the greenhouse environments to overcome cable wiring difficulties for the sensors and the LAN connection. Additionally, the flexibility in data sharing can be upgraded on the cloud system with user experience. The generated commands and decisions that are received by the IoT automation board from the cloud-based streaming system are used to control specific elements within the crop growth microenvironment while at the same time they can be optimized by the onboard computer to reduce energy, chemicals, and water demands. It can be concluded that developing a robust and affordable IoT automation system for greenhouse condition should take into account the correct selection and combination of the battery and charging units, the electronic housing box, connectors and plugs, data wire and cables, wireless antenna, and the modularity and compatibility of the package components. Results of experiments inside different greenhouses with high-density plants showed that the major disadvantage of wireless sensor nodes in real-time monitoring is the repeated loss of connection even in mesh applications. The water in the high amount of biomass of the plants damps the radio signals and avoids communication distances over long ranges. This can be solved by using different techniques (that sometimes involve a huge amount of effort), including antennas with cable for higher positions, higher mesh density, multiple gateway nodes, and higher output power. In general, it is a good practice to store all measurement data using devices that benefits from local memory. Therefore, the asynchronous readout is enabled for the user, and the data is not missed which an efficient practice for IoT is monitoring in large-scale commercial berry production. It is expected that this process embraces the uncertainties, especially in the remote areas, and consequently contributes to a higher yield with lesser inputs.

Acknowledgements

The authors acknowledge the funding support of Adaptive AgroTech Consultancy International for accelerating the research and development phase in prototyping the instruments and conducting the experiments. The technical and editorial assistant received from Adaptive AgroTech members, Peyman Majidi, Keyvan Majidi, Janet Ahmadi, Saeedeh Pourhanife, Batuhan Sakal, and Omid Raftari, as well as the suggestions and guidelines received from Dr. Volker Dworak, Dr. Jana Käthner, and Professor. Cornelia Weltzien of the Leibniz Institute for Agricultural Engineering and Bioeconomy are duly acknowledged.

Author details

Redmond R. Shamshiri[1], Ibrahim A. Hameed[2], Kelly R. Thorp[3], Siva K. Balasundram[4], Sanaz Shafian[5], Mohammad Fatemieh[6], Muhammad Sultan[7], Benjamin Mahns[1] and Saba Samiei[8]*

1 Leibniz Institute for Agricultural Engineering and Bioeconomy, Potsdam-Bornim, Germany

2 Department of ICT and Natural Sciences, Faculty of Information Technology and Electrical Engineering, NTNUÅlesund, Norway

3 United States Department of Agriculture, Agricultural Research Service, Maricopa, AZ, USA

4 Department of Agriculture Technology, Faculty of Agriculture, Universiti Putra Malaysia, Serdang, Selangor, Malaysia

5 School of Plant and Environmental Sciences, Virginia Polytechnic Institute and State University, Blacksburg, USA

6 Adaptive AgroTech Consultancy Int, Seaside, CA, USA

7 Department of Agricultural Engineering, Bahauddin Zakariya University, Multan, Pakistan

8 Comfort.AI, Blockhouse Bay, Auckland, New Zealand

*Address all correspondence to: saba@macso.ai

References

[1] R. R. Shamshiri *et al.*, "Advances in greenhouse automation and controlled environment agriculture: A transition to plant factories and urban agriculture," *Int. J. Agric. Biol. Eng.*, vol. 11, no. 1, 2018.

[2] R. R. Shamshiri *et al.*, "Model-based evaluation of greenhouse microclimate using IoT-Sensor data fusion for energy efficient crop production," *J. Clean. Prod.*, p. 121303, 2020.

[3] S. M. Rezvani *et al.*, "IoT-Based Sensor Data Fusion for Determining Optimality Degrees of Microclimate Parameters in Commercial Greenhouse Production of Tomato," *Sensors*, vol. 20, no. 22, p. 6474, 2020.

[4] C. Serôdio, J. Boaventura Cunha, R. Morais, C. Couto, and J. Monteiro, "A networked platform for agricultural management systems," *Comput. Electron. Agric.*, vol. 31, no. 1, pp. 75-90, 2001.

[5] K. P. Ferentinos, N. Katsoulas, A. Tzounis, T. Bartzanas, and C. Kittas, "Wireless sensor networks for greenhouse climate and plant condition assessment," *Biosyst. Eng.*, vol. 153, pp. 70-81, 2017.

[6] R. R. Shamshiri, J. W. Jones, K. R. Thorp, D. Ahmad, H. C. Man, and S. Taheri, "Review of optimum temperature, humidity, and vapour pressure deficit for microclimate evaluation and control in greenhouse cultivation of tomato: a review," *Int. Agrophysics*, vol. 32, no. 2, pp. 287-302, 2018.

[7] R. Shamshiri, H. Che Man, A. J. Zakaria, P. V. Beveren, W. I. Wan Ismail, and D. Ahmad, *Membership function model for defining optimality of vapor pressure deficit in closed-field cultivation of tomato*, vol. 1152. 2017.

[8] X. Bai, Z. Wang, L. Sheng, and Z. Wang, "Reliable data fusion of hierarchical wireless sensor networks with asynchronous measurement for greenhouse monitoring," *IEEE Trans. Control Syst. Technol.*, no. 99, pp. 1-11, 2018.

[9] D. Ma, N. Carpenter, H. Maki, T. U. Rehman, M. R. Tuinstra, and J. Jin, "Greenhouse environment modeling and simulation for microclimate control," *Comput. Electron. Agric.*, vol. 162, no. November 2018, pp. 134-142, 2019.

[10] M. C. Singh, J. P. Singh, and K. G. Singh, "Development of a microclimate model for prediction of temperatures inside a naturally ventilated greenhouse under cucumber crop in soilless media," *Comput. Electron. Agric.*, vol. 154, no. August, pp. 227-238, 2018.

[11] W. Xin, W. Yu, Z. Yuanyuan, N. Xindong, and W. Shumao, "Intelligent Gateway for Heterogeneous Networks Environment in Remote Monitoring of Greenhouse Facility Information Collection," *IFAC-PapersOnLine*, vol. 51, no. 17, pp. 217-222, 2018.

[12] M. Mizoguchi, T. Ito, A. Chusnul, S. Mitsuishi, and M. Akazawa, "Quasi real-time field network system for monitoring remote agricultural fields," in *SICE Annual Conference 2011*, 2011, pp. 1586-1589.

[13] *A. Prima et al.*, "Development of a remote environmental monitoring and control framework for tropical horticulture and verification of its validity under unstable network connection in rural area," *Comput. Electron. Agric.*, vol. 124, pp. 325-339, 2016.

[14] J. W. Jones, A. Kenig, and C. E. Vallejos, "Reduced state–variable tomato growth model," vol. 42, no. 1994, pp. 255-265, 1999.

[15] A. Pérez-González, O. Begovich-Mendoza, and J. Ruiz-León, "Modeling of a greenhouse prototype using PSO and

differential evolution algorithms based on a real-time LabView^{TM} application," *Appl. Soft Comput.*, vol. 62, pp. 86-100, 2018.

[16] J. Chen, Y. Ma, and Z. Pang, "A mathematical model of global solar radiation to select the optimal shape and orientation of the greenhouses in southern China," *Sol. Energy*, vol. 205, pp. 380-389, 2020.

[17] E. Romantchik, E. Ríos, E. Sánchez, I. López, and J. R. Sánchez, "Determination of energy to be supplied by photovoltaic systems for fan-pad systems in cooling process of greenhouses," *Appl. Therm. Eng.*, vol. 114, pp. 1161-1168, 2017.

[18] D. A. and H. C. M. Redmond R. Shamshiri, Muhammad Razif Mahadi, Kelly R. Thorp, Wan Ishak Wan Ismail, "Adaptive management framework for evaluating and adjusting microclimate parameters in tropical greenhouse crop production systems," *Plant Eng.*, 2017.

[19] R. Shamshiri, P. van Beveren, H. Che Man, and A. J. Zakaria, "Dynamic assessment of air temperature for tomato (*Lycopersicon esculentum* mill) cultivation in a naturally ventilated net-screen greenhouse under tropical lowlands climate," *J. Agric. Sci. Technol.*, vol. 19, no. 1, 2017.

[20] R. Shamshiri *et al.*, "Comparative evaluation of naturally ventilated screenhouse and evaporative cooled greenhouse based on optimal vapor pressure deficit," in *2016 American Society of Agricultural and Biological Engineers Annual International Meeting, ASABE 2016*, 2016.

[21] D. Shamshiri, R., Ismail, W. I. W., & bin Ahmad, "Experimental evaluation of air temperature, relative humidity and vapor pressure deficit in tropical lowland plant production environments," *Adv. Environ. Biol.*, 2014.

[22] R. Shamshiri, "Measuring optimality degrees of microclimate parameters in protected cultivation of tomato under tropical climate condition," *Meas. J. Int. Meas. Confed.*, vol. 106, 2017.

Chapter 2

Temperature and Humidity Control for the Next Generation Greenhouses: Overview of Desiccant and Evaporative Cooling Systems

Muhammad Sultan, Hadeed Ashraf, Takahiko Miyazaki, Redmond R. Shamshiri and Ibrahim A. Hameed

Abstract

Temperature and humidity control are crucial in next generation greenhouses. Plants require optimum temperature/humidity and vapor pressure deficit conditions inside the greenhouse for optimum yield. In this regard, an air-conditioning system could provide the required conditions in harsh climatic regions. In this study, the authors have summarized their published work on different desiccant and evaporative cooling options for greenhouse air-conditioning. The direct, indirect, and Maisotsenko cycle evaporative cooling systems, and multi-stage evaporative cooling systems have been summarized in this study. Different desiccant materials i.e., silica-gels, activated carbons (powder and fiber), polymer sorbents, and metal organic frameworks have also been summarized in this study along with different desiccant air-conditioning options. However, different high-performance zeolites and molecular sieves are extensively studied in literature. The authors conclude that solar operated desiccant based evaporative cooling systems could be an alternate option for next generation greenhouse air-conditioning.

Keywords: desiccant dehumidification, evaporative cooling, temperature and humidity control, next generation greenhouse

1. Introduction

Plants are highly sensitive to a specified temperature and humidity range inside a greenhouse. Higher than normal humidity level inside a greenhouse can be fatal for plant growth which leaves the plants open to fungus/pests' attacks, and causes dripping due to condensation. Humidity inside a greenhouse is contributed through photosynthesis and evapotranspiration. Photosynthesis is a natural phenomenon inside plants (which occurs due to chlorophyll) during daytime which results in carbohydrates in plants using carbon dioxide and photons. As a result of this process, plants' leaves generate water vapors into the surroundings contributing to the humidity of the air signified by Eq. (1) and **Figure 1**. Additionally, air temperature inside greenhouse is also impacted by incident solar radiations/sunlight.

$$2nCO_2 + 4n(H_2O)_{xylem} \overset{sunlight}{\rightarrow} 2(CH_2O)_n + 2nO_2\uparrow + 2n(H_2O)\uparrow$$

Furthermore, evapotranspiration or ET is combined process of evaporation and transpiration. Evaporation of water vapors occurs from the soil inside the greenhouse and transpiration of water vapors occurs from stomata (i.e., small openings underside the plant leaves). It concludes that moisture is added into the surrounding air inside the greenhouse at night (through evapotranspiration) as well as at daytime (through photosynthesis). **Figure 1a** and **Figure 1b** show the addition of moisture into the air through evapotranspiration and photosynthesis processes, respectively. Optimum growth and flowering stages of plants is highly impacted by the level of carbon dioxide (CO_2) and relative humidity (RH) inside a greenhouse [1, 2]. Photosynthesis process requires some amount of CO_2 from the air whereas the RH is dependent on the vapor pressure deficit (VPD) inside the greenhouse. The VPD inside the greenhouse is impacted by the plant growth stage, climatic conditions, temperature of plant leaf, and temperature of the inside air (i.e., microclimate) [3, 4]. **Figure 2** shows the impact of different humidity levels inside next generation greenhouses.

In this study, the authors summarize their previously published studies on temperature and humidity control options in next generation greenhouses. Optimum temperature/humidity levels, vapor pressure deficit (VPD) inside the greenhouse, evaporative cooling systems, desiccant air-conditioning systems and different desiccant materials for greenhouse air-conditioning are reviewed in this study.

1.1 Optimum temperature/humidity in greenhouse

As established earlier, optimum temperature/ humidity conditions are required inside the greenhouse optimum functioning of plants. Different plants require specific ranges of temperature and humidity at different stages of their growth. **Figure 3** shows a desiccant dehumidification-based temperature/humidity control system for agricultural greenhouse air-conditioning. **Figure 3** shows the psychrometric representation of the desiccant air-conditioning system for greenhouse air-conditioning. Outdoor air is cooled using direct evaporative cooling (which increases the humidity ratio of the air), further cooled using a sensible heat exchanger before passing through a low-grade heating source (i.e., solar thermal, or biogas), which passes through the desiccant material for regeneration purpose. Whereas on the other side, process air from the system outlet is used for greenhouse

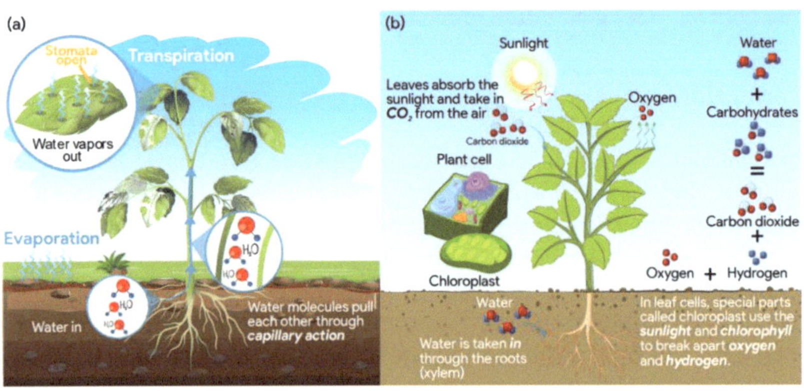

Figure 1.
Illustrations of (a) evapotranspiration (ET) process, and (b) photosynthesis process.

air-conditioning purpose on a simple recirculation mode using the MEC or IEC system for additional cooling.

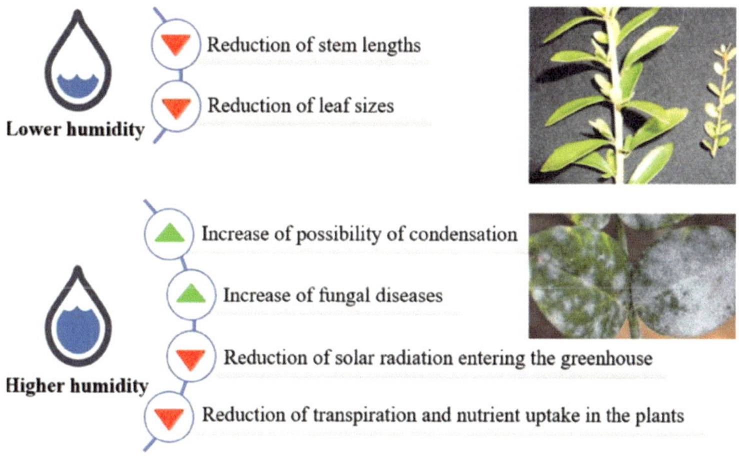

Figure 2.
Impact of different humidity level inside next generation greenhouses [5].

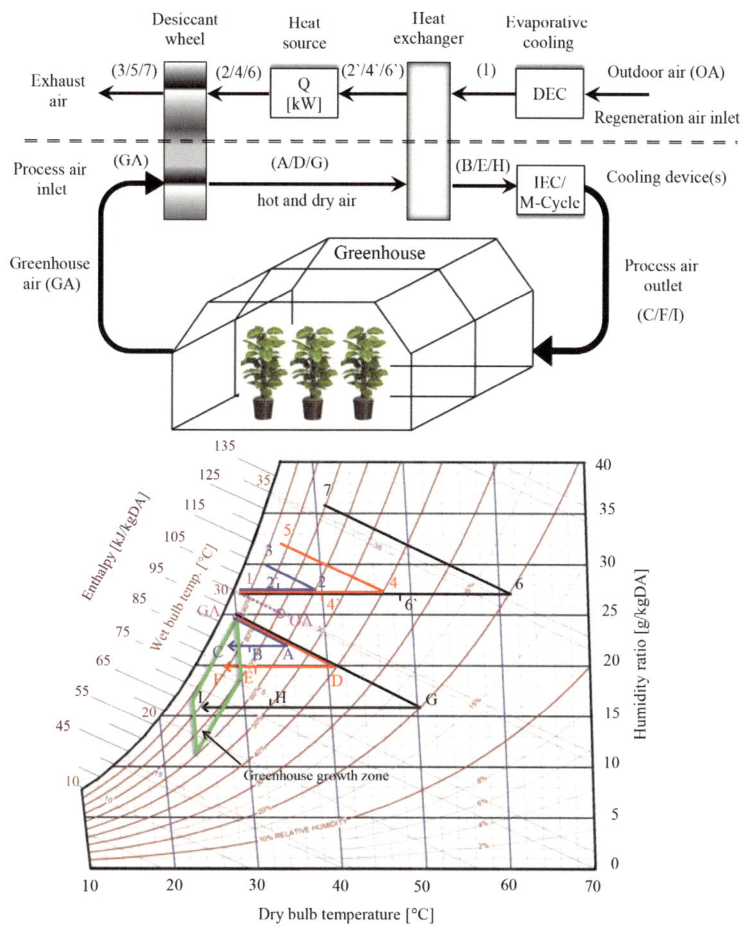

Figure 3.
(above) schematic representation of desiccant air-conditioning system, and (below) psychrometric working of the desiccant air-conditioning system, for next generation agricultural greenhouses [1].

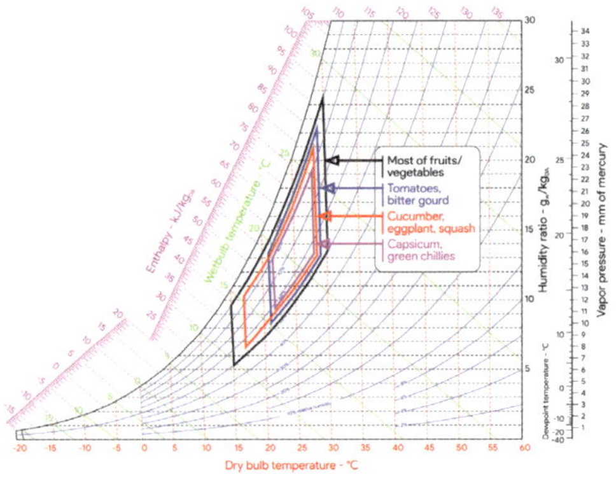

Figure 4.
Optimum temperature/humidity conditions for various fruits and vegetables inside an agricultural greenhouse.

Figure 4 shows the temperature and relative humidity range for optimum growth of various fruits and vegetable plants inside a typical next generation greenhouse. Most of the fruits and vegetable plants require 15 to 30°C temperature with 50 to 90% relative humidity for optimum growth inside an agricultural greenhouse. The most important parameter to address inside an agricultural greenhouse is vapor pressure deficit (VPD), which could be defined as a function of temperature of air, temperature of leaf, and relative humidity of the air. VPD changes with change in plant growth stage as well. Therefore, it is of paramount importance to address the vapor pressure deficit inside the greenhouse before designing a greenhouse air-conditioning system.

1.2 Vapor pressure deficit (VPD)

Vapor pressure deficit is a function of leaf temperature, air temperature, and relative humidity of the air. Vapor pressure deficit inside the greenhouse also varies with different growth stages, regions, and the plant being cultivated for crop. **Figure 5** shows the spatiotemporal profile of vapor pressure deficit across Pakistan. According to **Figure 5**, regions with higher relative humidity and lower temperature (i.e., Northern and North-Western regions of the country) tend to have relatively lesser vapor pressure deficit throughout the year. However, plain areas of the country with relatively higher temperature and drier climate tend to have peaks of vapor pressure deficit as high as 1.52 kPa for the month of June, which demands an air-conditioning system to regulate the temperature/humidity inside the greenhouse for optimum plant growth. The spatiotemporal profile (**Figure 5**) shows that air-conditioning is required inside an agricultural greenhouse in the plain areas (i.e., South, South-Eastern, and South-Western regions) of the country through April to September.

Additionally, **Figure 6** shows the correlation of vapor pressure deficit against dry bulb temperature and relative humidity of the air. According to **Figure 6**, most of the plants (fruits and vegetables) require a vapor pressure deficit of 0.45 kPa to 1.25 kPa for their normal growth throughout their different growth stages. However, for optimum growth of most fruits and vegetables, plants require a vapor pressure deficit of 0.8 to 0.9 kPa. **Figure 7** shows the experimental results of vapor pressure deficit of a next generation agricultural greenhouse. According to **Figure 7**, the vapor pressure deficit inside an agricultural greenhouse was comparatively higher than the outside conditions throughout the day. The study concluded that the developed evaporative fan/pad cooling systems were most feasible for regions with relative humidity lesser than 65% from the viewpoint of vapor pressure deficit.

Moreover, studies show that fungi diseases survive below 0.4 kPa vapor pressure deficit therefore an appropriate air-conditioning system is required for maintaining

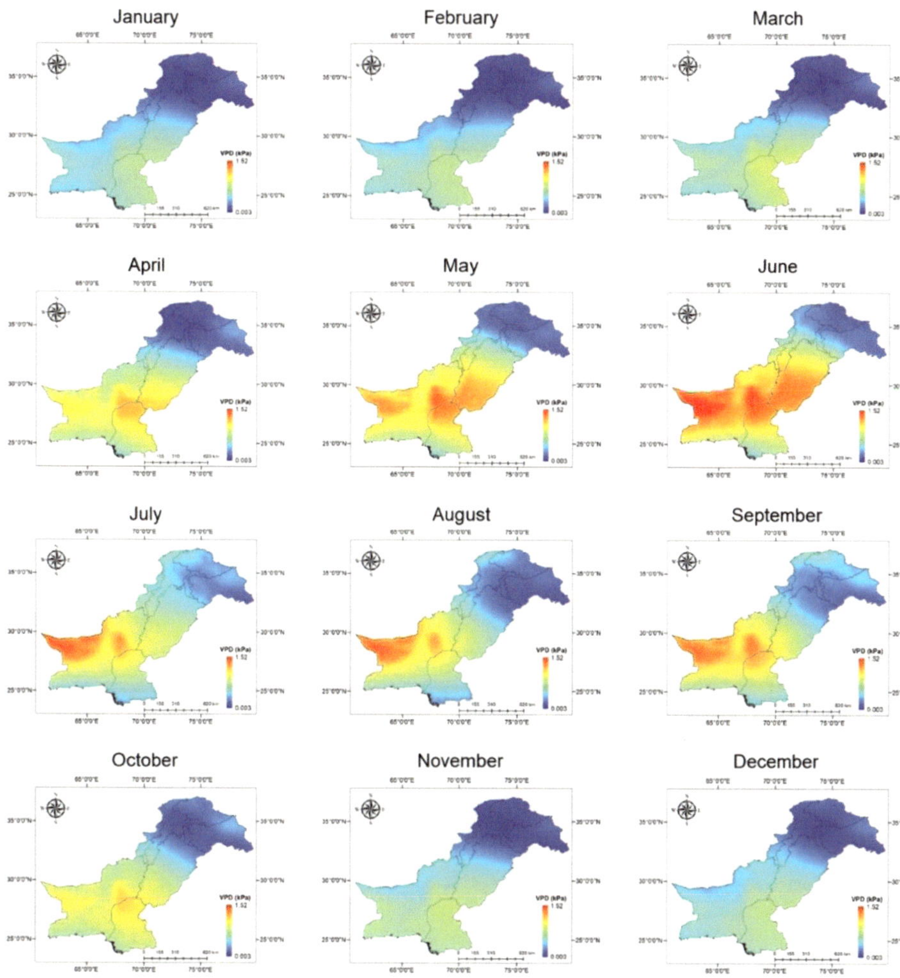

Figure 5.
Spatiotemporal profile of vapor pressure deficit (VPD) across Pakistan [6].

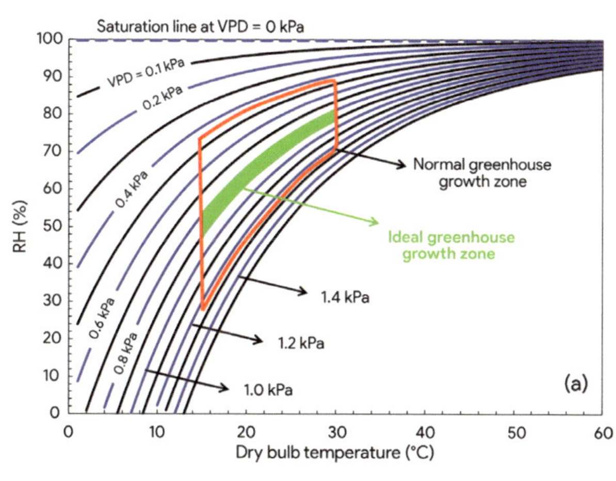

Figure 6.
Identifying the normal and ideal greenhouse growth zones for agricultural products on the basis of water vapor pressure deficit, reproduced from [1].

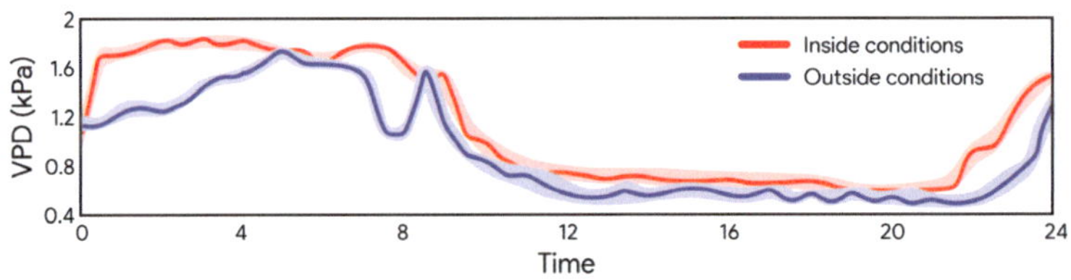

Figure 7.
Vapor pressure deficit (VPD) inside a next generation greenhouse for tomato case study, reproduced from [7].

Optimum ranges			Marginal ranges			Ref.
VPD (kPa)	T (°C)	RH (%)	VPD (kPa)	T (°C)	RH (%)	
0.75–1.06	15–34	40–85	0.45–1.25	15–34	35–85	[2]
0.5–1.2	15–35	35–90	0.4–1.37	17–34	35–90	[8]
0.4–0.79	15–30	55–90	—	—	—	[9]
0.47–1.27	15–30	60–85	—	—	—	[10]

Table 1.
Optimum and marginal ranges of vapor pressure deficit (VPD), temperature (T), and relative humidity (RH) inside next generation greenhouses.

the humidity inside greenhouses. **Table 1** shows optimum and marginal ranges of vapor pressure deficit, temperature, and relative humidity inside next generation greenhouses from different sources. Some of the case studies (presented in Section 1.3) show the variation in vapor pressure deficit, air temperature, relative humidity and solar irradiance being received for different crops at different growth stages.

1.3 Case studies

Numerous next generation greenhouses have been studied for different plant/crops under controlled climate environment (i.e., microclimate) in literature. Shamshiri R. R. et al. [4] evaluated IoT sensors inside two different types (i.e., screenhouse and Polycarbonate sheet greenhouse) next generation greenhouses for energy efficient crop production of tomato plant. The authors studied different growth stage requirements of tomato crop and developed a novel comfort ratio model (i.e., *Cft* model) which was further validated using MATLAB Simulink. **Figure 8** shows the dynamic inputs/outputs of the study on tomato crop.

Moreover, **Figure 9** shows the difference in air temperature, relative humidity, vapor pressure deficit, and irradiance, between the two studied greenhouses (i.e., naturally ventilated screenhouse and energy efficient Polycarbonate sheet evaporative cooling-based greenhouse). The results indicate that the comfort ratio index (which takes into account dynamic assessment of each input parameter i.e., air temperature, relative humidity, vapor pressure deficit, and irradiance, at different timeframes) could be more insightful parameter for comparing the energy efficient crop production between any two studied greenhouses. **Figure 9** shows the raw data used for the IoT sensor-based modeling for two next greenhouses. From the studied literature, it can be concluded that typically naturally ventilated screenhouses and evaporative cooling-based energy efficient greenhouses are in practice for enhanced crop production. However, evaporative cooling systems fail to perform optimally under certain high humidity conditions. In this regard, thermally driven dehumidification based evaporative cooling systems could be an option for additional humidity control in such regions.

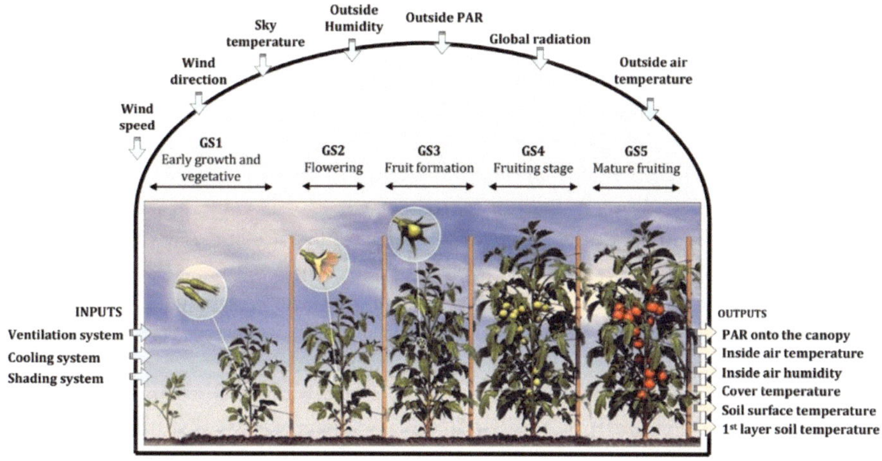

Figure 8.
Illustration of stages of tomato growth inside next generation greenhouses [4].

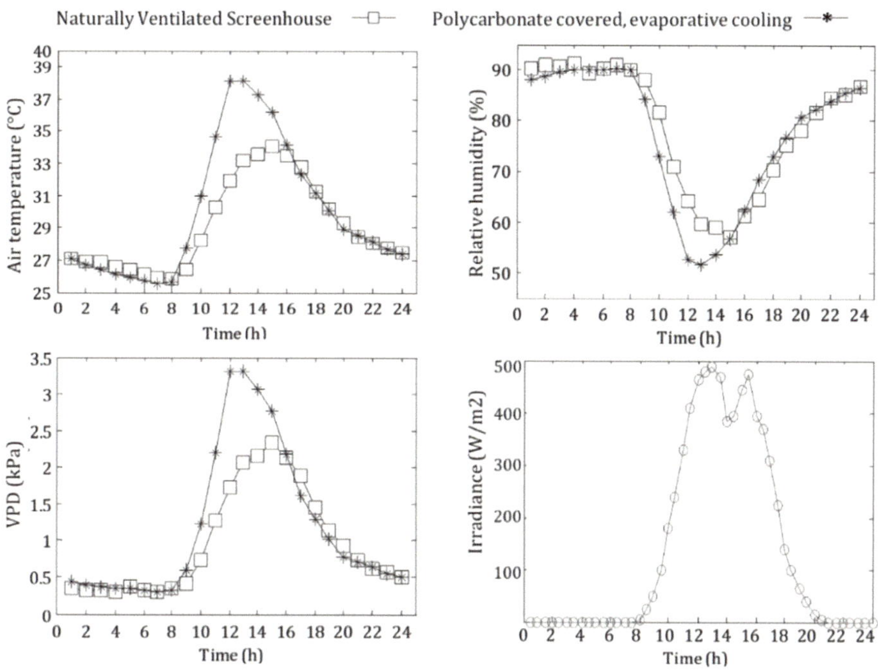

Figure 9.
Experimental results profile of the microclimate inside a next generation greenhouse [4].

2. Evaporative cooling systems

2.1 Evaporative cooling options

Evaporative cooling (EC) is a traditional cooling technique which utilizes endothermic energy from the phase change of water from liquid into water vapors. This phase change mostly occurs due to a current of air using the heat energy present in air for the phase change. Typically, regions with higher temperatures, evaporative cooling is preferred as a low-cost option for air-conditioning, however it does have its drawbacks. Too high temperature can sometimes result in failure of EC systems.

Moreover, humidity present in air also presents a limitation on the performance of the EC systems. The EC systems are, as a thumb rule, suitable for regions with high temperatures and relatively lesser humidity in the air [11]. The authors have thermodynamically, experimentally, and numerically investigated mainly three types of EC systems (i.e., direct evaporative cooling (DEC), indirect evaporative cooling (IEC), and Maisotsenko cycle evaporative cooling (MEC)) for various climatic conditions and air-conditioning applications [11–15]. In direct evaporative cooling (DEC) system, inlet air is in direct contact with wet channels (wet medium usually water with honeycomb/*khas*) which causes cooling due to evaporation but also increases the humidity of the ambient/inlet air at the outlet. On the other hand, in the IEC system, inlet air is in indirect contact with wet channels (wet medium usually water with aluminum channel walls). It causes cooling due to evaporation in the wet channel, conducts the cooling through the channel walls into the inlet air flowing in dry channel. Both the DEC and IEC systems are psychrometrically limited to the wet bulb temperature of the inlet/ambient air however, the MEC can create a temperature gradient of below wet bulb temperature up to dewpoint temperature of the inlet/ambient air. The MEC system is an advanced form of the IEC system with slight modification. Some portion of the outlet air from dry channel is diverted into the wet channel which ultimately results in temperature gradient lower than wet bulb temperature of the ambient/inlet air. **Figure 10(a)** shows the schematic diagram and psychrometric working principle of traditional MEC system. Another modification in the traditional MEC system is shown in **Figure 10(b)**. Mahmood et al. studied the MEC system for various air-conditioning applications [16]. **Figure 11** shows the experimental setup of the developed IEC and MEC systems. The authors concluded that the standalone MEC system was only able to achieve the required temperature/ humidity conditions when the humidity ratio of the ambient air was ≤11 g/kgDA. In other words, in humid regions with humidity ratio higher than 11 g/kgDA, the standalone MEC system failed to produce the optimum temperature/ humidity conditions as the temperature increased. Higher humidity ratio in the air resulted in relatively higher temperature gradient (i.e., according to **Figure 12**, the

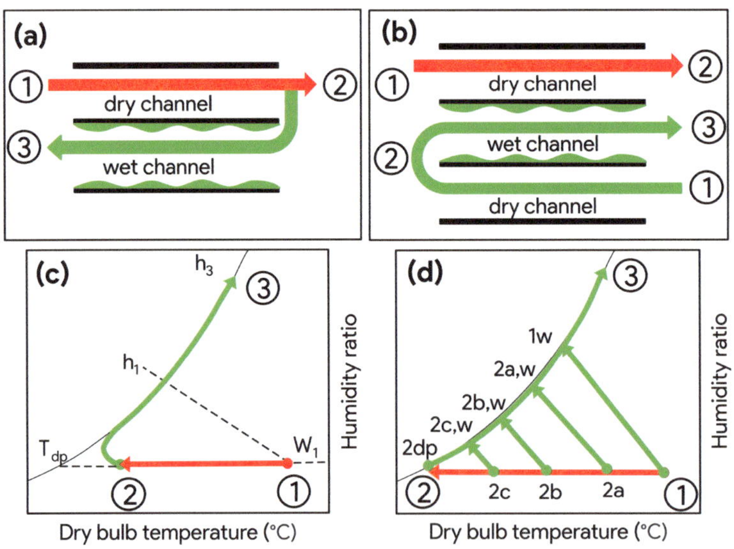

Figure 10.
Schematic illustrations (a,b) and psychrometric representations (c,d) of traditional Maisotsenko cycle evaporative cooling (MEC) system, and modified Maisotsenko cycle evaporative cooling system, respectively, reproduced from [16].

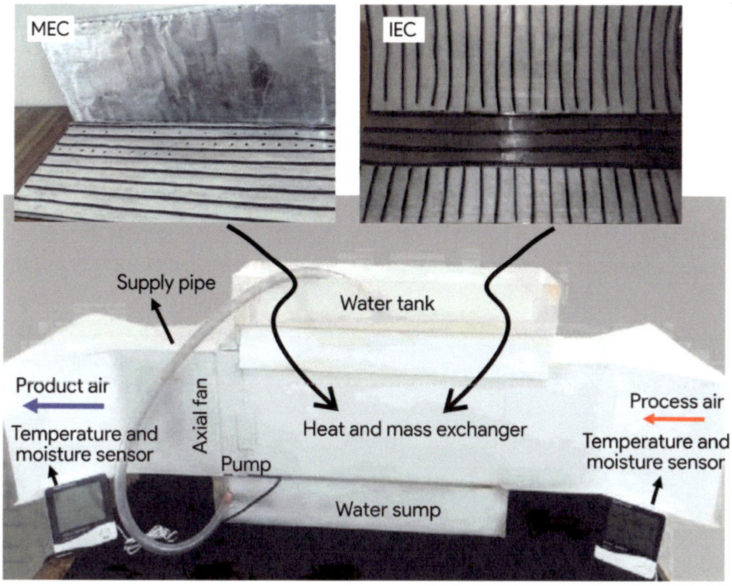

Figure 11.
Experimental setup of the IEC and MEC systems, developed by the authors, reproduced from [17].

MEC system cooled the ambient air with 25 g/kgDA humidity down to 27°C, whereas 20°C in case of 11.2 g/kgDA humidity ratio).

In conclusion, the DEC system creates an addition of moisture into the supply air; increasing its humidity ratio in an uncontrolled fashion which psychrometrically limits the evaporative cooling potential, is undesirable in next generation greenhouses. Whereas the IEC system offers a solution for the uncontrolled humidity problem faced in the DEC system. However, the IEC system is psychrometrically limited to the wet bulb temperature of the ambient air which results in lower temperature gradient, unsuitable for next generation greenhouses. In this regard, an advanced IEC system or the MEC system proves to be effective which is theoretically psychrometrically limited to the dewpoint temperature of the ambient air which results in relatively more temperature gradient, suitable for next generation greenhouses. However, the MEC system is not feasible for regions where humidity in air is ≤11 g/kgDA. In such regions, multi-staging of evaporative cooling systems could an appropriate option for various air-conditioning applications.

2.2 Multi-staging of evaporative cooling systems

Psychrometrics limits the evaporative cooling of the air to its wet bulb temperature (in case of direct evaporative cooling) and dewpoint temperature (in case of Maisotsenko cycle evaporative cooling). However, multi-staging of evaporative cooling systems with other types of evaporative cooling systems or vapor compression air-conditioning (VCAC) systems can break above mentioned conventional bonds of psychrometrics and achieve below wet bulb temperature and effectiveness. Authors have previously worked on multi-staging numerical analysis of evaporative cooling coupled with vapor compression air-conditioning systems and their possible combinations. Noor et al. studied DEC, IEC, and VCAC systems and their possible combinations for building air-conditioning and concluded that multi-stage IEC-DEC-VCAC achieved maximum temperature gradient of ~21°C and wet bulb effectiveness of 1.4 for the summer conditions of Multan [11]. **Figure 12** shows the illustrations and

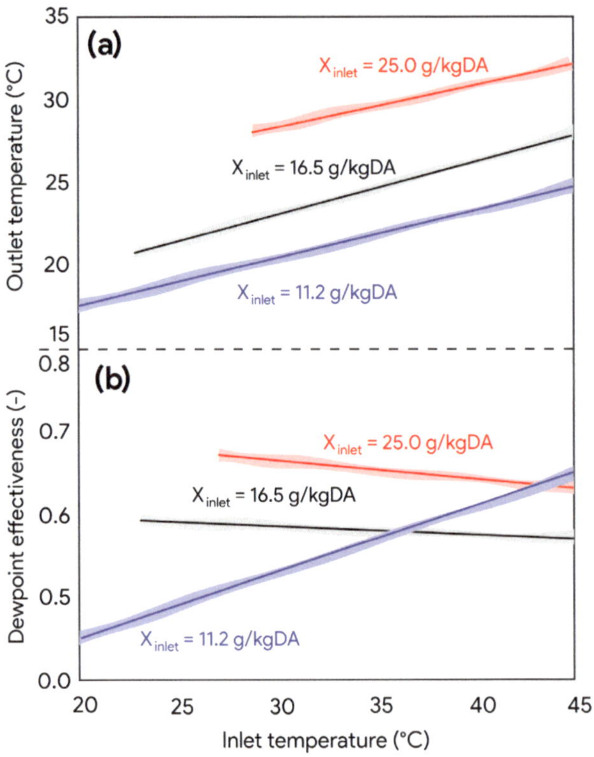

Figure 12.
Effect of inlet temperature on (a) outlet temperature, reproduced from [18], and (b) dewpoint effectiveness of MEC, reproduced from [18].

psychrometric representations of multi-staging of DEC, IEC, and VCAC systems for building air-conditioning application. **Figure 13** shows the temperature gradient profile of the DEC, IEC, and VCAC systems and their possible combinations. As can be seen from **Figure 13**, the IEC-DEC-VCAC created a maximum temperature gradient (i.e., ~21°C), whereas the IEC system created the least temperature gradient (i.e., ~10°C) for summer conditions of Multan (Pakistan). Moreover, the IEC-DEC-VCAC system consumed relatively less power as compared to standalone VCAC system. Additionally, the IEC-DEC-VCAC system had a carbon dioxide release equivalent of 241,134 kg/CO_2/year which is relatively lesser than the standalone VCAC system (i.e., 274,883 kg/CO_2/year). Moreover, the IEC-DEC-VCAC created maximum cooling capacity (i.e., 184 kW) and its work input was ~100 kW. However, maximum COP of the IEC-DEC-VCAC was 2.1, which is relatively lesser as compared to the standalone DEC system (i.e., 4.5) due to higher work input at VCAC stage. Thus, the authors concluded that the IEC-DEC-VCAC system could potentially achieve the desired conditions of any environment including building air-conditioning and other air-conditioning applications at a relatively lower cost (**Figure 14**).

2.3 Evaporative cooling for greenhouse temperature/humidity control

Evaporative cooling could be an energy efficient option for greenhouse air-conditioning in regions where ambient conditions are not feasible for naturally ventilated screenhouses. Noor et al. [6] investigated the thermodynamic performance of direct, indirect and Maisotsenko cycle evaporative cooling systems for greenhouse air-conditioning under the climatic conditions of Multan (Pakistan). The authors took into account the air temperature, relative humidity, vapor pressure deficit, and wet

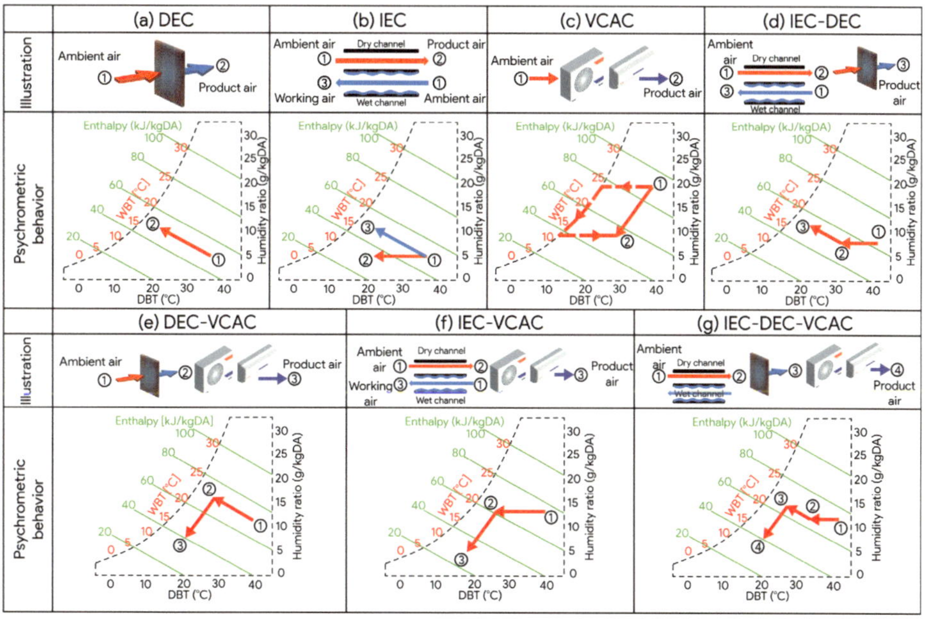

Figure 13.
Schematic illustration and psychrometric working of all possible multi-stages of the direct evaporative cooling (DEC), indirect evaporative cooling (IEC), and vapor compression air-conditioning (VCAC) systems [11].

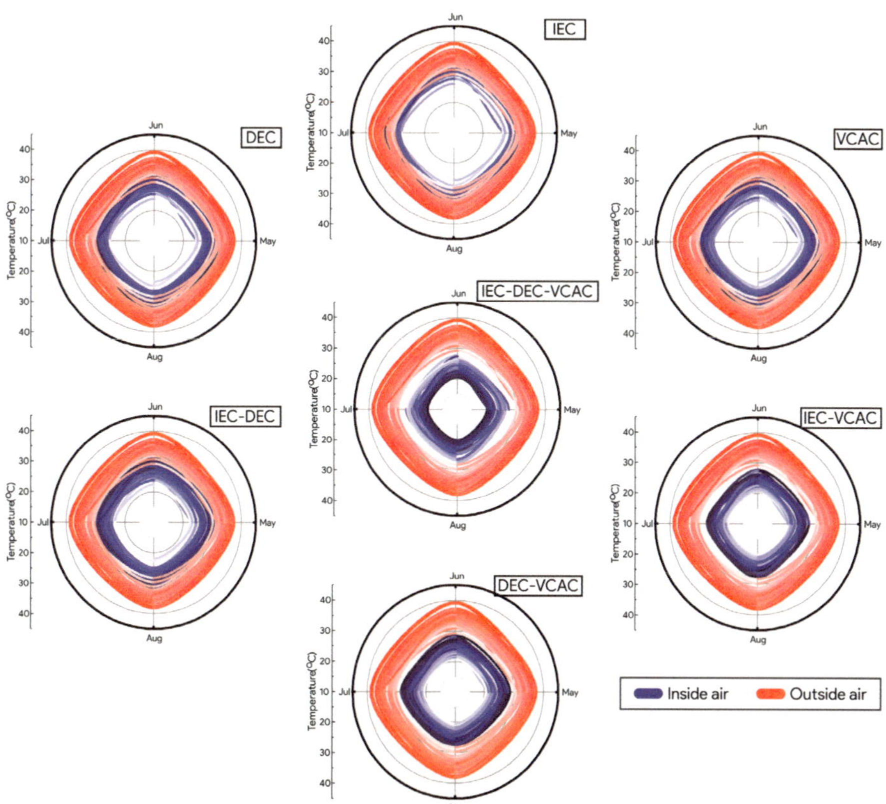

Figure 14.
Temperature gradient profile of the DEC, IEC, and VCAC systems and their possible combinations [11].

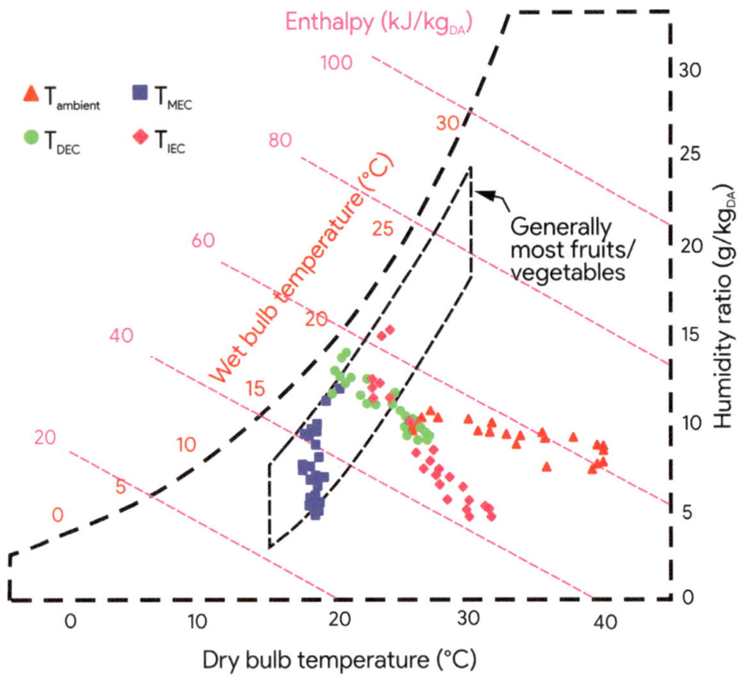

Figure 15.
Psychrometric performance of direct (DEC), indirect (IEC), and Maisotsenko (MEC) cycle evaporative cooling systems for greenhouse air-conditioning (15th may) for Multan (Pakistan), reproduced from [6].

bulb and dewpoint effectiveness of the systems. The authors concluded that the ambient conditions of the study area were not feasible for greenhouse temperature/ humidity requirements, and only the MEC system were able to achieve the desired temperature/ humidity conditions. Moreover, the DEC system was partially able to achieve the required temperature/humidity conditions inside next generation greenhouse. However, the IEC system failed to achieve the required conditions. The authors also concluded that the DEC system achieved the maximum wet bulb effectiveness of 0.9, whereas the MEC system was able to achieve maximum dewpoint effectiveness of 0.5 to 0.6. However, the DEC system was not feasible for greenhouse air-conditioning due to high humidity ratio in the product air throughout the year which indicates that more feasible energy efficient option could be explored (**Figure 15**).

In this regard, the MEC system could be a suitable option for greenhouse air-conditioning. However, performance of the MEC system is also limited to ambient air-conditions (see Section 2.1 in this Chapter, **Figure 12**) which essentially indicates that MEC system fails to perform in too high humidity conditions. Therefore, desiccant dehumidification based evaporative cooling options could be an alternative option for next generation greenhouse air-conditioning in regions with higher humidity ratio.

3. Desiccant air-conditioning systems

3.1 Desiccant materials

3.1.1 Silica-gels

Desiccant air-conditioning has been extensively studied in literature for various applications including next generation greenhouse air-conditioning. Performance of any desiccant air-conditioning system is still somewhat limited to the surrounding conditions however these systems performs admirably well in humid climates unlike the conventional evaporative cooling system which cap at wet bulb temperature of the ambient air. Performance of the desiccant air-conditioning system varies with the desiccant material being used for adsorption of moisture and regeneration temperature for desorption of the material. Silica-gels serve as a viable, cost effective solution for moisture uptake as opposed to expensive, high efficiency experimental polymers, activated carbons, metal organic frameworks, zeolites, and molecular sieves. Sultan et al. [1] investigated the experimental performance of silica-gel based desiccant air-conditioning system for greenhouse air-conditioning at different regeneration temperatures. The experimental results were validated through modeling. **Figure 16** shows the adsorption uptake performance of the silica-gel at 20, 30, and 50°C regeneration temperatures against pressure. According to **Figure 16**, silica-gel adsorption uptake is relatively equal at lower regeneration temperature and pressure as compared to high temperature and pressure, which makes it economically feasible and viable in regions with harsh conditions. However, the underlying fact cannot be denied that performance of other desiccant materials (discussed in coming Sections) is relatively higher than the silica-gels.

3.1.2 Activated carbons

Activated carbons (i.e., powder (ACP) and fiber (ACF)) relatively adsorb higher amount of moisture as compared to conventionally used, easily available silica-gel. Sultan et al. [1] experimentally investigated the adsorption uptake onto the activated carbon powder (Maxsorb-III) and activated carbon fiber (A-20) for next generation solar operated greenhouse air-conditioning. The aim of the study was to investigate the performance of different desiccant materials at different regeneration temperatures. The authors concluded that at 30°C regeneration temperature, the activated carbon powder adsorption uptake was 3 times higher as compared to conventionally used silica-gel whereas it was 1.5 times higher in case of activated carbon fiber vs. silica-gel. **Figure 17** shows the adsorption uptake performance of the activated carbon powder (Maxsorb-III) at 20, 30, and 50°C against different pressures. According to **Figure 17**, the adsorption uptake in case of ACP at 50°C regeneration temperature was 1.4 kg/kg whereas it was 1.25 kg/kg and 1.2 kg/kg at 20 and 30°C, respectively.

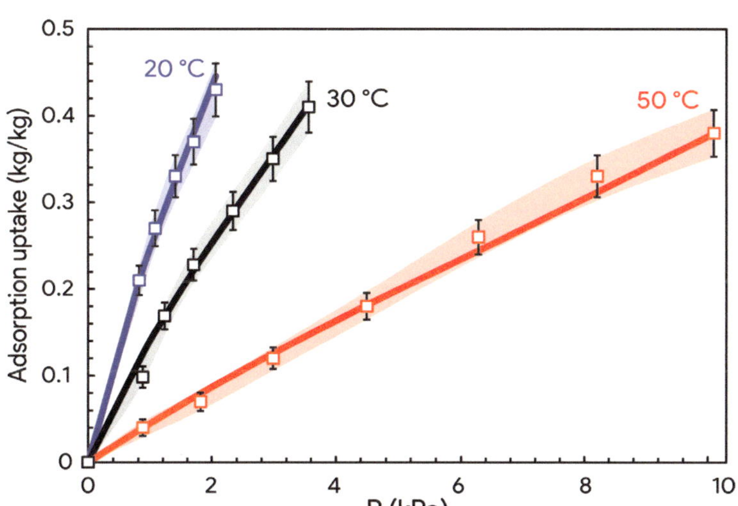

Figure 16.
Adsorption of water vapor onto silica-gel at different temperature, reproduced from [1].

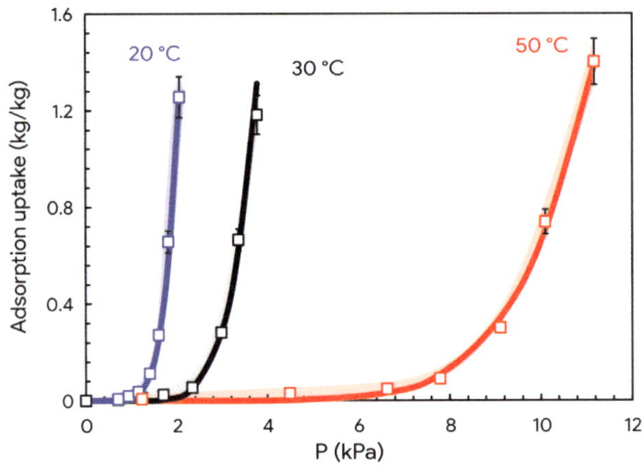

Figure 17.
Adsorption of water vapor onto activated carbon powder (ACP) at different temperature, reproduced from [1].

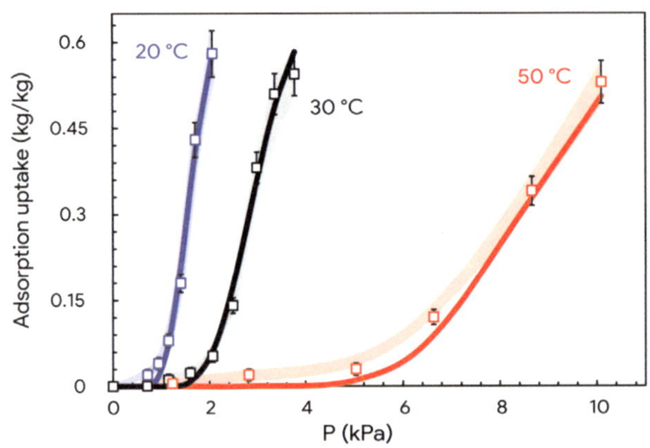

Figure 18.
Adsorption of water vapor onto activated carbon fiber (ACF) at different temperature, reproduced from [1].

Whereas **Figure 18** shows the adsorption uptake performance of the activated carbon fiber (A-20) at 20, 30, and 50°C against pressure. According to **Figure 18**, the adsorption uptake in case of ACF at 50°C regeneration temperature was 0.47 kg/kg whereas it was 0.6 kg/kg and 0.5 kg/kg at 20 and 30°C, respectively.

Figure 19 shows the Polanyi's adsorption potential of the activated carbon powder and activated carbon fiber against conventionally used silica-gel. According to **Figure 19**, the activated carbon powder and activated carbon fiber have higher adsorption uptake as compared to conventionally used silica-gel when the adsorption uptake potential is lower than 50 kJ/kg, which could be considered as the threshold limit for greenhouse air-conditioning [1]. **Figure 20** shows the mass fraction of the adsorbent material per mass of air required for dehumidifying the air. According to **Figure 20**, the activated carbon powder was able to produce the best results for demand category-I with the least amount of mass fraction of adsorbent required. All three materials, silica-gel, ACP, and ACF were able to satisfy the demand category-I whereas only ACF and the conventionally used silica-gel was able to achieve the required output conditions of demand category-II. However, neither ACF nor ACP were able to satisfy the demand category-III which has the maximum humidity gradient (i.e., difference between inlet and outlet humidity) which makes silica-gel more suitable in case of demand category-III.

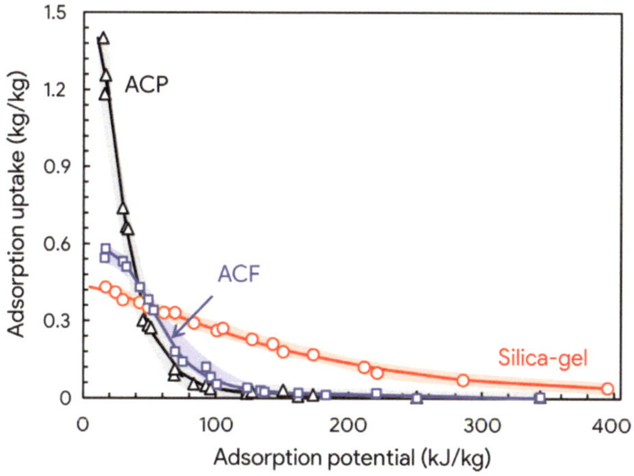

Figure 19.
Profile of adsorption uptake equilibrium against different adsorption potential for silica gel, ACP, and ACF. Points: Experiment; lines: General trend, reproduced from [1].

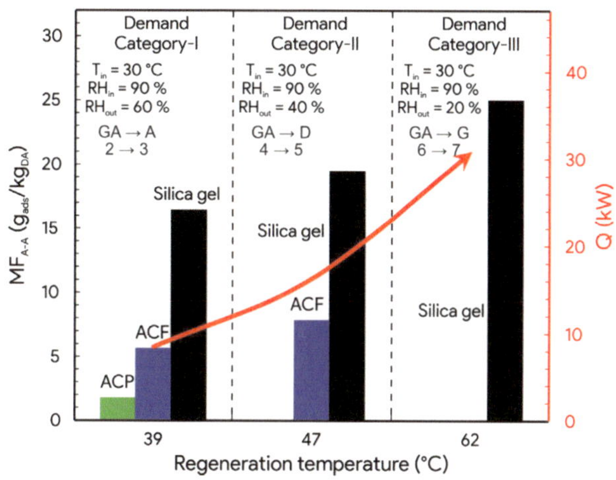

Figure 20.
Profile of regeneration temperature and heating energy using silica-gel, ACP and ACF for different next generation greenhouses' humidity demand, reproduced from [1].

3.1.3 Polymer sorbents

In this study, experimental performance of two polymeric sorbents i.e., PS-I and PS-II has been investigated from authors' previous work [19]. Sultan et al. [19] investigated the experimental performance of moisture uptake onto the PS-I/water and PS-II/water pairs for desiccant air-conditioning applications. **Figure 21(a)** shows the adsorption uptake isotherms of PS-I and PS-II at 30°C adsorption temperature. According to **Figure 21(a)**, PS-II type polymeric sorbent has higher adsorption uptake (i.e., 0.9 kg/kg) at 30°C adsorption temperature whereas it was 0.6 in case of PS-I. **Figure 21(b)** shows the adsorption kinetics of the PS-I and PS-II water adsorption uptake. The authors concluded that PS-I was able to achieve higher rate of adsorption uptake/dehumidification at relatively low regeneration temperature (i.e., 50°C) as compared to PS-II. Although PS-II was able to produce better steady state adsorption kinetics, however it did not have an impact on the overall performance of the system.

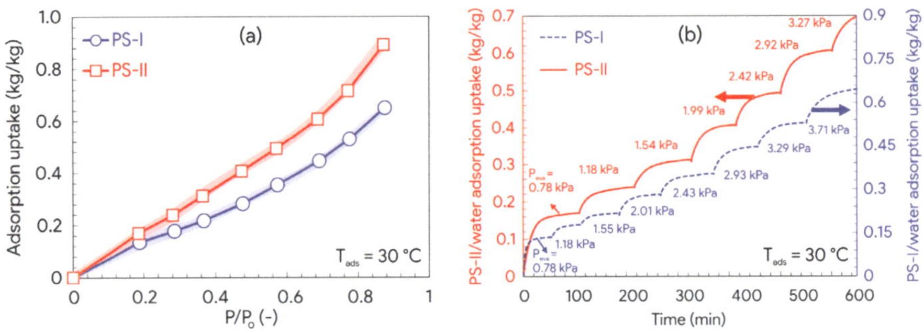

Figure 21.
Performance profile of PS-I and PS-II at T_{ads} = 30, (a) isotherms for PS-I and PS-II, points: Experiments; lines: General trend; fill area: Experimental uncertainty, reproduced from [20], and (b) adsorption kinetics for PS-I and PS-II, reproduced from [20].

3.1.4 Metal organic frameworks (MOFs)

Figure 22 shows the adsorption uptake performance of four different metal organic frameworks reviewed from literature. According to **Figure 22(a)** and **(c)**, MIL-101(Cr) and HKUST-1 produced the highest adsorption uptake i.e., 1.45 kg/kg and 0.50 kg/kg, at 25°C adsorption temperature, respectively. Whereas CPO-27(Ni) and MIL-53 produced the highest adsorption uptake i.e., 0.45 kg/kg and 0.90 kg/kg, at 30°C adsorption temperature, respectively. HKUST-1 shows more degree of stability at low pressure. Whereas adsorption uptake performance of the MIL-53 was negatively impacted by increase in adsorption temperature. However, CPO-25 (Ni) produced relatively same adsorption uptake performance throughout the variation in pressure due to all particle sizes.

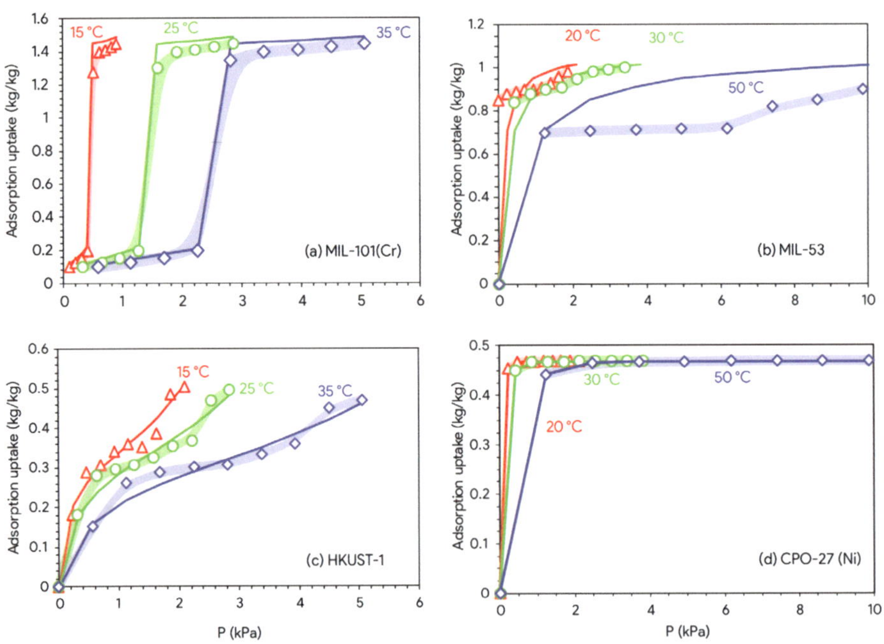

Figure 22.
Adsorption uptake for (a) MIL-101(Cr) [21, 22], (b) MIL-53 [23, 24], (c) HKUST-1 [25, 26], and (d) CPO-27(Ni) [27, 28], reproduced from [29].

3.1.5 Others

Apart from the silica-gels, activated carbons (powder or otherwise fiber), polymer sorbents, and metal organic frameworks, different types of zeolites, and molecular sieves have been investigated in literature whose adsorption uptake performance relative to the studied materials could be higher but for the sake of simplicity and ease, only few materials i.e., silica-gels, activated carbons, polymer sorbents, and metal organic frameworks have been accounted for in this study.

3.2 Desiccant air-conditioning options

Desiccant air-conditioning for different air-conditioning options has been extensively studied in literature. Niaz et al. [30] investigated silica-gel desiccant dehumidification based Maisotsenko cycle evaporative cooling system for livestock thermal comfort. **Figure 23** shows the performance of the studied standalone desiccant and evaporative cooling based desiccant air-conditioning options for livestock thermal comfort for climatic conditions of Multan (Pakistan). The authors concluded that only the desiccant dehumidification based Maisotsenko cycle evaporative cooling system was able to achieve the desired temperature/humidity conditions for livestock thermal comfort. Whereas the ambient and standalone desiccant air-conditioning system conditions were unfavorable for livestock thermal comfort. Moreover, Aleem et al. [31] investigated the experimental performance of sieve/layers type orientation of silica-gel beads and compared it with polymeric sorbents. The authors concluded that polymeric sorbent had higher adsorption uptake as compared to the silica-gel. **Figure 24** shows the experimental setup, and the schematic diagram of the developed desiccant air-conditioning system.

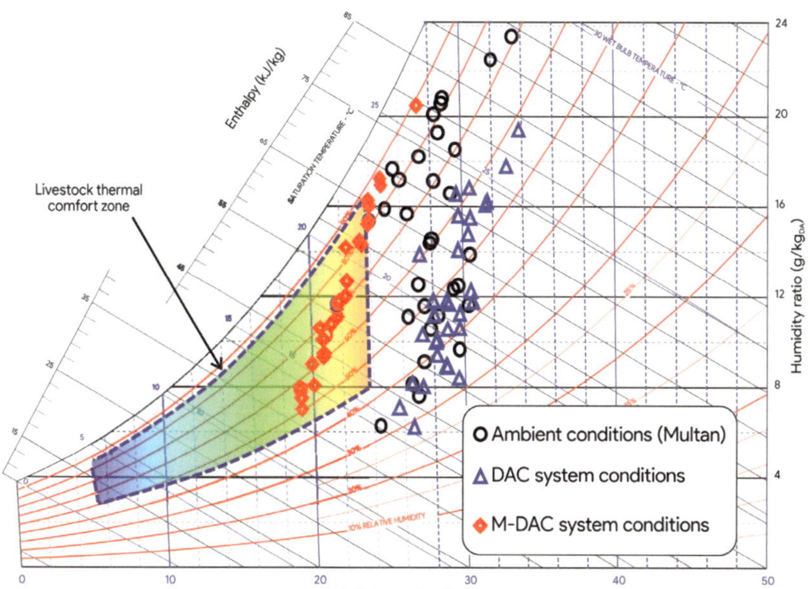

Figure 23.
Thermodynamic performance of desiccant dehumidification based Maisotsenko cycle evaporative cooling system for livestock thermal comfort for Multan (Pakistan), reproduced from [30].

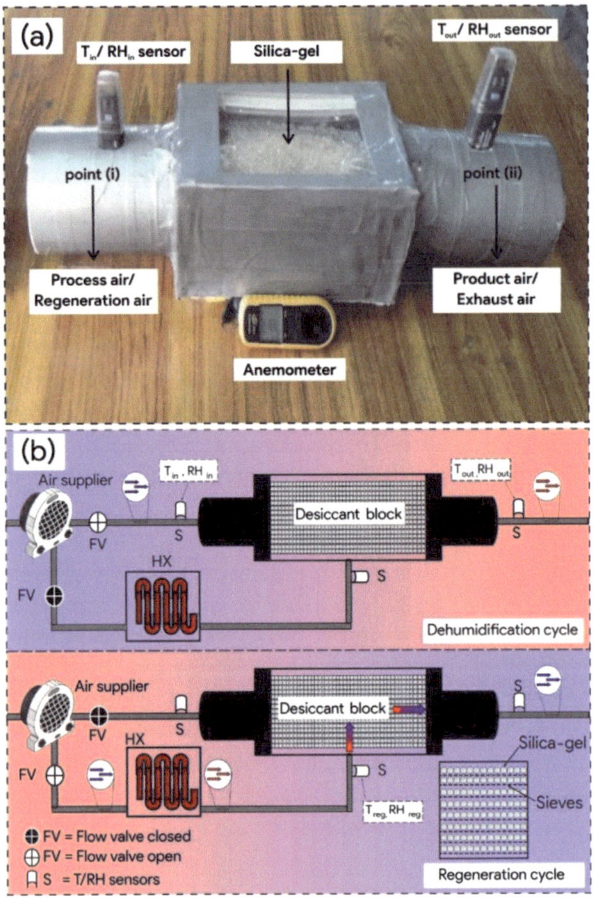

Figure 24.
Illustration of the experimental silica-gel desiccant dehumidification system, (a) snapshot of the developed system, and (b) schematic illustration of the developed system, reproduced from [31].

3.3 Desiccant air-conditioning for greenhouse temperature/humidity control

Authors have extensively studied desiccant air-conditioning options for the next generation greenhouse air-conditioning in literature [1, 3, 4, 14, 32–35]. Ashraf et al. [33] studied silica-gel desiccant dehumidification based Maisotsenko cycle evaporative cooling system for the next generation agricultural greenhouse. **Figure 25(a)** shows the schematic diagram of the investigated desiccant air-conditioning system. **Figure 25(b)** shows the annual performance profile of the desiccant dehumidification based Maisotsenko cycle evaporative cooling system for the climatic conditions of Multan (Pakistan) for next generation greenhouse air-conditioning. According to **Figure 25(b)**, the evaporative cooling enhanced desiccant air-conditioning system achieved the required conditions of greenhouse air-conditioning in the lower limit. However, the standalone desiccant and the ambient conditions were not feasible through most of the year for next generation greenhouse. The authors concluded that solar operated silica-gel desiccant based Maisotsenko cycle evaporative cooling system could be an alternate low cost, energy efficient, feasible solution for greenhouse air-conditioning.

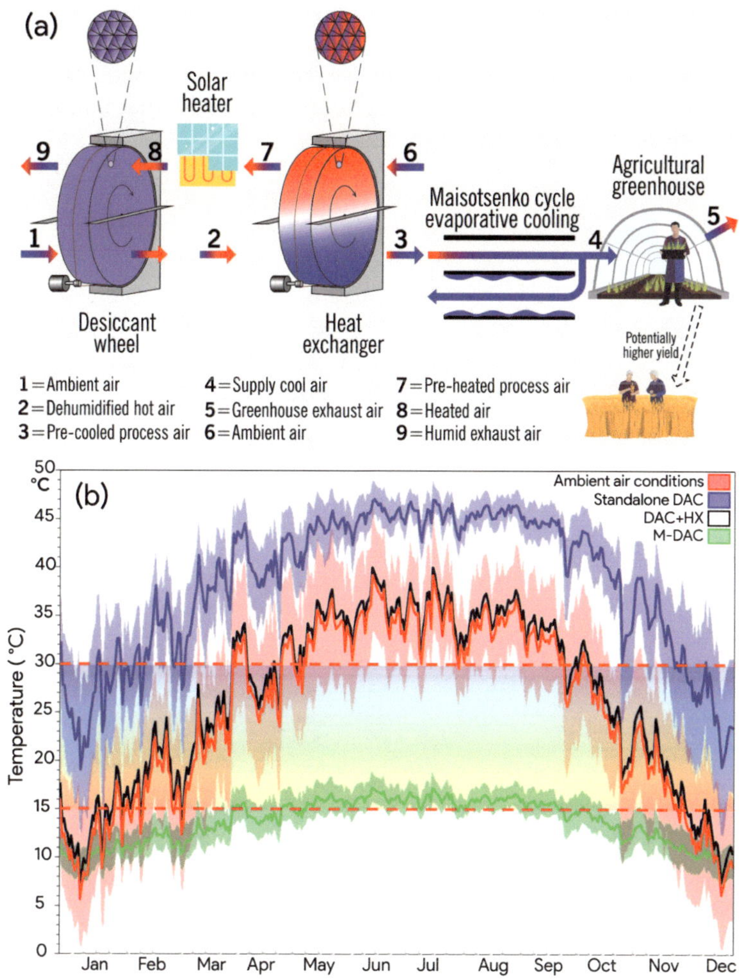

Figure 25.
(a) Schematic diagram of desiccant based Maisotsenko cycle evaporative cooling system, and (b) performance profile of standalone DAC and M-DAC systems for greenhouse air-conditioning in Multan (Pakistan), reproduced from [33].

4. Conclusions and summary

In this study, the authors summarized different evaporative cooling and desiccant air-conditioning options for next generation greenhouses. The authors summarized direct, indirect, and Maisotsenko cycle evaporative cooling options for different air-conditioning options including next generation greenhouses from the point of view of optimum temperature/humidity conditions and vapor pressure deficit inside the greenhouse. The authors concluded that only Maisotsenko cycle evaporative cooling system was able to achieve the required conditions under limited climatic applicability. However, performance of the Maisotsenko cycle evaporative cooling system could be enhanced using desiccant dehumidification. The authors summarized different desiccant materials (i.e., silica-gels, activated carbons, polymer sorbents, and metal organic frameworks) and impact of desiccant material and adsorption temperature on adsorption uptake. The authors concluded that each of the studied desiccant material somewhat achieved the required conditions. However, polymer sorbents and metal organic frameworks had relatively the highest adsorption uptake as compared to other materials. However, silica-gel based desiccant dehumidification system was able to achieve relatively more humidity gradient (i.e., difference between inlet and outlet humidity) for more adsorbent material to air mass fraction.

Acknowledgements

This work is carried out in the Department of Agricultural Engineering, Bahauddin Zakariya University, Multan-Pakistan. The work is financially supported by the Director Research/ORIC grants of Bahauddin Zakariya University (Multan), awarded to Principal Investigator Dr. Muhammad Sultan.

Conflict of interest

The authors declare no conflict of interest.

Author details

Muhammad Sultan[1*†], Hadeed Ashraf[1†], Takahiko Miyazaki[2,3], Redmond R. Shamshiri[4] and Ibrahim A. Hameed[5]

1 Department of Agricultural Engineering, Bahauddin Zakariya University, Multan, Pakistan

2 Faculty of Engineering Sciences, Kyushu University, Fukuoka, Japan

3 International Institute for Carbon-Neutral Energy Research (WPI-I2CNER), Kyushu University, Fukuoka, Japan

4 Department of Engineering for Crop Production, Leibniz Institute for Agricultural Engineering and Bioeconomy, Potsdam-Bornim, Germany

5 Department of ICT and Natural Sciences, Faculty of Information Technology and Electrical Engineering, Norwegian University of Science and Technology (NTNU), Ålesund, Norway

*Address all correspondence to: muhammadsultan@bzu.edu.pk

† These authors contributed equally to this work.

References

[1] Sultan M, Miyazaki T, Saha BB, Koyama S. Steady-state investigation of water vapor adsorption for thermally driven adsorption based greenhouse air-conditioning system. Renew Energy. 2016;86:785–95. https://doi.org/10.1016/j.renene.2015.09.015

[2] Shamshiri RR, Jones JW, Thorp KR, Ahmad D, Man HC, Taheri S. Review of optimum temperature, humidity, and vapour pressure deficit for microclimate evaluation and control in greenhouse cultivation of tomato: a review. Int Agrophysics. 2018;32(2):287–302. https://doi.org/10.1515/intag-2017-0005

[3] Sultan M, Miyazaki T, Koyama S, Saha BB. Utilization of Desiccant Air-Conditioning System for Improvement in Greenhouse Productivity: A Neglected Area of Research in Pakistan. Int J Environ. 2014;4(1):1–10.

[4] Shamshiri RR, Bojic I, van Henten E, Balasundram SK, Dworak V, Sultan M, et al. Model-based evaluation of greenhouse microclimate using IoT-Sensor data fusion for energy efficient crop production. J Clean Prod. 2020;263:121303. https://doi.org/10.1016/j.jclepro.2020.121303

[5] Amani M, Foroushani S, Sultan M, Bahrami M. Comprehensive review on dehumidification strategies for agricultural greenhouse applications. Appl Therm Eng. 2020;181:115979. https://doi.org/10.1016/j.applthermaleng.2020.115979

[6] Noor S, Ashraf H, Hussain G, Sultan M, Miyazaki T, Shakoor A, et al. Spatiotemporal investigation of evaporative cooling options for greenhouse air-conditioning application in Pakistan. Fresenius Environ Bull. 2021;21(2).

[7] Shamshiri R, Ismail WIW. Performance evaluation of ventilation and pad-and-fan systems for greenhouse production of tomato in lowland Malaysia. World Res J Agric Biosyst Eng. 2012;1(1):1–5.

[8] Argus. Understanding and Using VPD [Internet]. White Rock, Canada; 2009. Available from: http://www.arguscontrols.com/resources/VPD_Application_Note.pdf

[9] Omafra S. Growing Greenhouse Vegetables. Guelph, Ontario, Canada: Ontario Ministry of Agriculture. Food, Rural Aff. 2005;

[10] Ministry of Agriculture. Understanding Humidity Control in Greenhouses [Internet]. British Columbia; 2015. Available from: https://www2.gov.bc.ca/assets/gov/farming-natural-resources-and-industry/agriculture-and-seafood/animal-and-crops/crop-production/understanding_humidity_control.pdf

[11] Noor S, Ashraf H, Sultan M, Khan ZM. Evaporative Cooling Options for Building Air-Conditioning: A Comprehensive Study for Climatic Conditions of Multan (Pakistan). Energies. 2020 Jun 12;13(12):3061. https://doi.org/10.3390/en13123061

[12] Noor S, Ashraf H, Sultan M, Miyazaki T, Mahmood MH, Khan ZM. Investigation of Direct and Indirect Evaporative Cooling Options for Greenhouse Air Conditioning in Multan (Pakistan). In: Proceedings of International Exchange and Innovation Conference on Engineering & Sciences (IEICES). Kyushu University: Interdisciplinary Graduate School of Engineering Sciences, Kyushu University; 2020. p. 110–5. http://hdl.handle.net/2324/4102474

[13] Mahmood MH, Sultan M, Miyazaki T. Solid desiccant dehumidification-based

air-conditioning system for agricultural storage application: Theory and experiments. Proc Inst Mech Eng Part A J Power Energy. 2020;234(4):534–47. https://doi.org/10.1177/0957650919869503

[14] Sultan M, Miyazaki T. Energy-Efficient Air-Conditioning Systems for Nonhuman Applications. In: Ekren O, editor. Refrigeration. InTechOpen; 2017. p. 97–117.

[15] Sultan M, Niaz H, Miyazaki T. Investigation of Desiccant and Evaporative Cooling Systems for Animal Air-Conditioning. Low-temperature Technologies. Kyushu University, Japan; 2020.

[16] Mahmood MH, Sultan M, Miyazaki T, Koyama S, Maisotsenko VS. Overview of the Maisotsenko cycle – A way towards dew point evaporative cooling. Renew Sustain Energy Rev. 2016;66:537–55. https://doi.org/10.1016/j.rser.2016.08.022

[17] Raza HMU, Ashraf H, Shahzad K, Sultan M, Miyazaki T, Usman M, et al. Investigating Applicability of Evaporative Cooling Systems for Thermal Comfort of Poultry Birds in Pakistan. Appl Sci. 2020;10(13):4445. https://doi.org/10.3390/app10134445

[18] Sultan M, Miyazaki T, Mahmood MH, Khan ZM. Solar assisted evaporative cooling based passive air-conditioning system for agricultural and livestock applications. J Eng Sci Technol. 2018 Mar 1;13(3):693–703.

[19] Sultan M, Miyazaki T, Koyama S, Khan ZM. Performance evaluation of hydrophilic organic polymer sorbents for desiccant air-conditioning applications. Adsorpt Sci Technol. 2018; 36(1–2):311–26. https://doi.org/10.1177/0263617417692338

[20] Sultan M. Study on sorption characteristics of water adsorbents for agricultural air-conditioning systems. Kyushu University, Japan; 2015.

[21] Khutia A, Rammelberg HU, Schmidt T, Henninger S, Janiak C. Water sorption cycle measurements on functionalized MIL-101Cr for heat transformation application. Chem Mater. 2013;25(5):790–8.

[22] D'Ans P, Courbon E, Permyakova A, Nouar F, Simonnet-Jégat C, Bourdreux F, et al. A new strontium bromide MOF composite with improved performance for solar energy storage application. J Energy Storage. 2019;25:100881. https://doi.org/10.1016/j.est.2019.100881

[23] Teo HWB, Chakraborty A. Water adsorption on various metal organic framework. In: IOP Conference Series: Materials Science and Engineering. IOP Publishing; 2017. p. 12019.

[24] Jeremias F, Fröhlich D, Janiak C, Henninger SK. Advancement of sorption-based heat transformation by a metal coating of highly-stable, hydrophilic aluminium fumarate MOF. RSC Adv. 2014;4(46):24073–82.

[25] Rezk A, Al-Dadah R, Mahmoud S, Elsayed A. Experimental investigation of metal organic frameworks characteristics for water adsorption chillers. Proc Inst Mech Eng Part C J Mech Eng Sci. 2012 Aug 9;227(5):992–1005. https://doi.org/10.1177/0954406212456469

[26] Zhao Z, Wang S, Yang Y, Li X, Li J, Li Z. Competitive adsorption and selectivity of benzene and water vapor on the microporous metal organic frameworks (HKUST-1). Chem Eng J. 2015;259:79–89. https://doi.org/10.1016/j.cej.2014.08.012

[27] Youssef PG, Dakkama H, Mahmoud SM, AL-Dadah RK. Experimental investigation of adsorption water desalination/cooling

system using CPO-27Ni MOF. Desalination. 2017;404:192–9. https://doi.org/10.1016/j.desal.2016.11.008

[28] Jiao Y, Morelock CR, Burtch NC, Mounfield III WP, Hungerford JT, Walton KS. Tuning the kinetic water stability and adsorption interactions of Mg-MOF-74 by partial substitution with Co or Ni. Ind Eng Chem Res. 2015;54 (49):12408–14. https://doi.org/10.1021/acs.iecr.5b03843

[29] Ashraf S, Sultan M, Hussain G, Noor S, Ashraf H, Miyazaki T, et al. Investigation of water-vapor adsorption onto metal-organic frameworks (MOFs) for greenhouse air-conditioning application. Fresenius Environ Bull. 2021;

[30] Niaz H, Sultan M, Khan AA, Miyazaki T, Feng Y, Khan ZM, et al. Study on evaporative cooling assisted desiccant air conditioning system for livestock application in Pakistan. Fresenius Environ Bull. 2019;28(11 A): 8623–33.

[31] Aleem M, Hussain G, Sultan M, Miyazaki T, Mahmood MH, Sabir MI, et al. Experimental Investigation of Desiccant Dehumidification Cooling System for Climatic Conditions of Multan (Pakistan). Vol. 13, Energies. 2020. https://doi.org/10.3390/en13215530

[32] Sultan M, Miyazaki T, Saha BB, Koyama S, Maisotsenko VS. Steady-state Analysis on Thermally Driven Adsorption Air-conditioning System for Agricultural Greenhouses. In: Procedia Engineering. Elsevier B.V.; 2015. p. 185–92. https://doi.org/10.1016/j.proeng.2015.08.417

[33] Ashraf H, Sultan M, Shamshiri RR, Abbas F, Farooq M, Sajjad U, et al. Dynamic Evaluation of Desiccant Dehumidification Evaporative Cooling Options for Greenhouse Air-Conditioning Application in Multan (Pakistan). Energies. 2021 Feb 19;14(4): 1097. https://doi.org/10.3390/en14041097

[34] Sultan M, Niaz H, Miyazaki T. Investigation of Desiccant and Evaporative Cooling Systems for Animal Air-Conditioning. In: Refrigeration and Air-Conditioning. IntechOpen; 2019.

[35] Shahzad K, Sultan M, Bilal M, Ashraf H, Farooq M, Miyazaki T, et al. Experiments on Energy-Efficient Evaporative Cooling Systems for Poultry Farm Application in Multan (Pakistan). Sustainability. 2021 Mar 5;13 (5):2836. https://doi.org/10.3390/su13052836

Chapter 3

The Canadian Integrated Northern Greenhouse: A Hybrid Solution for Food Security

David Leroux and Mark Lefsrud

Abstract

Food security has become a prominent issue in northern Canada. Many constraints, including environmental, cultural and economic barriers to cause food insecurity in northern Canada where local food production is one proposed solution to the northern food crisis. Initiated at McGill University by the Biomass Production Laboratory, the Canadian Integrative Northern Greenhouse (CING) unit provides a completely integrative design solution that could allow northern Canadian communities to grow their own fresh and nutritious food year-round. The CING unit is a hybrid between a northern greenhouse and a growth chamber housed in a shipping container, designed to be adaptive by functioning as a typical solar greenhouse when solar light provides considerable heat and light, and as a closed growth chamber during the night and when colder, darker winter conditions prevail. The CING was designed and prototyped by McGill students since 2013. Lettuce was grown during the four-season test of the CING, the greatest yield obtained was in March 2019, where the plants grown achieved 72% of the dry mass of the plants grown in the research greenhouse. The CING relied on supplemental heating to successfully grow plants but demonstrated the potential for northern and remote applications.

Keywords: shipping container farming, controlled environment agriculture, northern agriculture, northern greenhouse, organic fertilizer

1. Introduction

The CING is designed as a hybrid between a closed growth chamber and a greenhouse to optimize energy requirements related to the production of fresh produce throughout the year. The unit can open to allow sunlight to enter, utilizing the unit's greenhouse function, or be completely covered by an insulated thermal curtain, employing the unit's growth chamber function. Specific exterior and interior conditions dictate when the use of each mode is most efficient to promote the best interior conditions. To determine and predict these conditions, climatic and environmental data were recorded outside and inside the CING prototype situated at McGill University's Macdonald Campus in Sainte-Anne-de-Bellevue, QC, since summer 2015.

IntechOpen

1.1 Container farming

Container farming (CF) is an indoor agricultural practice falling under the Controlled Environment Agriculture (CEA) category [1]. Plants are grown hydroponically in a shipping container with electrical lighting and most of the environmental parameters are controlled by the grower. Converting a shipping container into an indoor farm has many advantages. First, a shipping container is an inexpensive infrastructure. Buying a refurbished shipping container and modifying its structure by cutting through the walls is still considered cheaper than buying a new building. Second, transportation, if the structural components of a shipping container are intact (i.e. the four corner beams), the CF has a strong foundation that can be moved as a typical shipping container. In this way, it acts as a mobile agricultural unit. Third, a converted shipping container's internal environment is independent of environmental parameters. In an insulated environment comprising electrical lighting, soil-less cultures, and heating ventilating and air conditioning (HVAC) technologies, it is possible to grow crops in any climate. Finally, a converted shipping container offers a high yield per square meter. Using vertical farming in which five levels of shallow water hydroponic cultures of lettuce are stacked, it is possible to grow 20 times more produce per square meter in a CF than field agriculture with corresponding yields of 1000 plants. m^{-2} [2].

CF is still a relatively new agricultural practice, and indoor farmers do not necessarily agree that this new agricultural practice is economically viable, still being considered an overhyped technology, with only 50% of container farms being profitable in the U.S. [3]. Yet CF has many different styles, with companies such as Freight Farms, Growtainers, and Cubic Farms offering similar options to grow crops in urban or remote areas [4] (**Figure 1**). According to case studies from companies like Bright Agrotech and independent reports from universities such as the University of Bonn in Germany and the Massachusetts Institute of Technology, vertical farming and CF can be economically profitable and viable depending on different economic parameters, such as market, labor and cheap energy availability [5].

The concept of a modified shipping container for controlled environment agriculture is not new (**Figure 2**). Strategies using modified shipping containers with natural lighting has been made for conditions comparable to those found in New York City and Los Angeles by the University of Arizona. From these simulations, it was determined that shipping containers with transparent walls have a much lower energy consumption than opaque and well-insulated walls (**Table 1**) [6].

Figure 1.
Outside of the CING, December 2017.

Figure 2.
A module for the Minimally Structured & Modular Vertical Farm, designed by Dr. Cuello from the University of Arizona (Liu, 2014).

Annual Energy Estimation (kWh/m²)	Los Angeles		New York City	
	Transparent wall	Opaque wall	Transparent wall	Opaque wall
Tomato	240.06	381.30	557.65	325.34
Lettuce	418.38	1950.99	773.84	1640.85

Table 1.
Summary of annual energy consumption in kWh/m2 [6].

From these energy values, except for growing tomatoes in a transparent wall shipping container in New York City where the well-insulated opaque wall helped reduce heat loss in a colder month, using transparent walls in a shipping container would reduce the energy needs to grow certain food crops in CF, even for Lettuce during cold months [6]. Following these findings, the CING was not modeled for its energy use, rather, design and experimental approach was chosen to test the use of natural lighting in CF in a cold climate.

2. Materials and methods

2.1 Design of the CING

The CING was first designed in 2013 by Bioresource engineering students at McGill University (**Figures 3** and **4**). A shipping container was purchased in 2015.

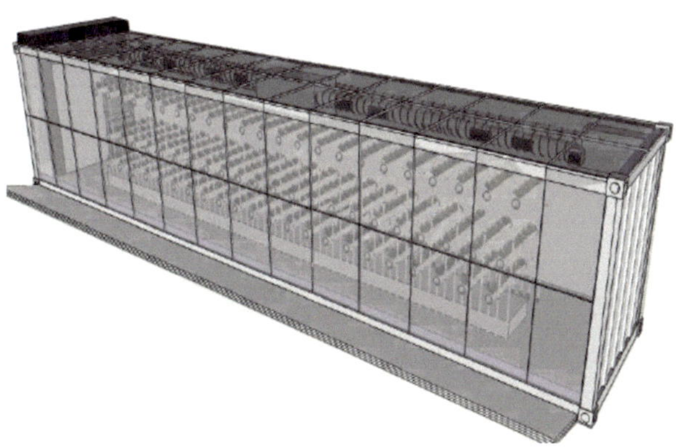

Figure 3.
Original design of the CING.

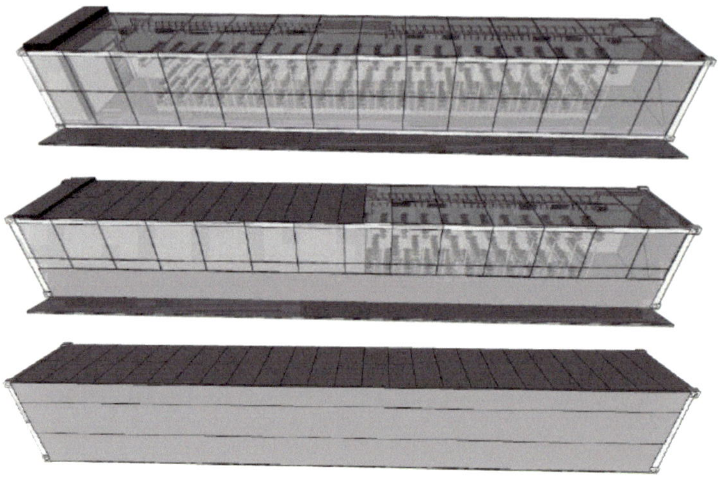

Figure 4.
Representation of the opening and closing of the outside panels.

One of its walls and the roof were replaced by polycarbonate sheets to allow the shipping container to use natural light for growing purposes.

Only half of the 40-foot shipping container was used for growing space. The CING design includes insulating panels that can open and close (added in 2015) to benefit from natural light when available (**Figures 5** and **6**). Their opening and closing were operated by 2000-lb winches controlled by an Arduino Mega (Adafruit Industries, US).

A growth tower was designed to allow inter-canopy lighting of the crops, optimizing the use of the supplemental electrical light. The growth tower was originally designed for drip irrigation (**Figure** 7).

In 2017, the tower was converted to a nutrient film technique (NFT). A comparative tower was built using a similar inter-canopy pattern for testing the CING's performance which was placed in a research greenhouse at McGill University's Macdonald Campus (**Figure 8**).

Figure 5.
Opening (left) and closing (right) of the CING insulating panels.

Figure 6.
Opening (right) and closing (left) of the CING rooftop panels.

Figure 7.
Original design of the CING growth tower (left), side-view (top right) and solution tank (bottom right), pictures by Thanh Jutras, 2016.

2.2 Energy usage

One of the CING operational challenge was using minimal energy consumption. It was determined that the CING must be operational on a 30-Amp, 110 V-circuit year-round, for maximum daily energy usage of 79.2 kWh (**Table 2**) (Eq. 1).

$$\text{Energy (kWh)} = \text{Current(A)} * \text{Voltage (V)} * \text{time (h)}/1000 \quad (1)$$

For this reason, supplemental lighting and heating are limited, but the use of natural light as a light and heat source for the growing environment was the main parameter studied to evaluate the CING's potential as an energy-efficient indoor growing system adapted for a northern climate.

Under cold weather conditions, the exhaust fans were not used while in warm weather the heaters were not used resulting in maximum daily energy uses of 29.4

Figure 8.
Comparative growth tower in the research greenhouse, Summer 2018.

Equipment	AC Current (amps)	Voltage (V)
Irrigation pump (4 pumps)	3.2	110
Heaters	13.8	110
LED lights	3.3	110
Automation control system	1*	110
Motor for thermal curtains	1*	110
Exhaust fans	2.12	110
Total	24.42	110

**The estimated current was required for automation system and thermal curtains function.*

Table 2.
Electrical current and voltage consumption of the CING environment control system components [7].

kWh.m^{-2} and 14.0 kWh.m^{-2} respectively. These values were obtained using only a small, representative growing area of 2 m^2. The growing area of half of 40-foot shipping is 14.4 m^2. More lighting, pumping capacity and air exchange would be needed if the full growing area was used.

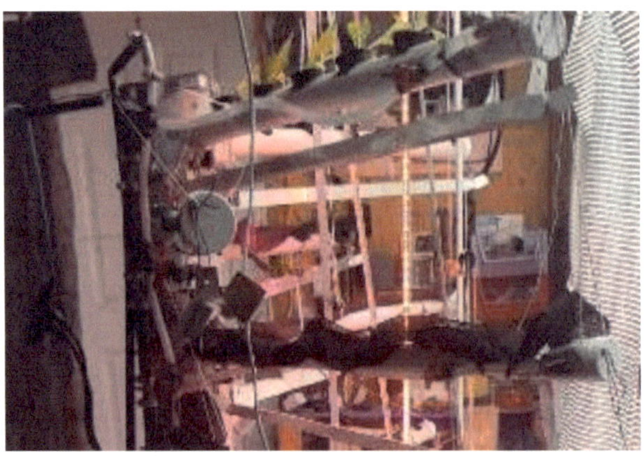

Figure 9.
Inside the CING, on the right is the closed thermal curtain, Winter 2019.

2.3 Thermal curtain parameters

A thermal curtain (TEMPA 7567 D FB, Svensson, North Carolina, U.S.), allowed a transition from greenhouse mode to growth chamber mode (**Figure 9**). The thermal curtain was functional and set to open when solar irradiation was above 12 W.m^{-2} and close when irradiation went lower than the set value. This value was recommended in a previous report on the recommended operating conditions of the CING [7].

2.4 Growth experiments

The CING ran for four consecutive seasons: Spring 2018 (May 7th to June 6th), Summer 2018 (June 8th to July 2nd), Fall 2018 (December 1st to December 22nd) and Winter 2019 (March 1st to March 23rd).

2.5 Biological nutrient solution testing

Since both growing systems had two independent pumps for the right and left sides, two nutrient solutions were tested in each system. The first was a one-quarter strength Hoagland solution [8] and the second comprised a biological nutrient solution based on vermicompost leachate. This solution was continuously prepared during the experiment using 10 L vermicompost, fed a constant diet of eggshells, banana peels, coffee grounds and cardboard. By flooding the vermicompost weekly with 1 L water, the leachate was collected and diluted to match the electrical conductivity (EC) of the Hoagland nutrient solution.

2.6 Hydroponic systems parameters

2.6.1 Design

The hydroponic growth systems were built as growing towers (**Figures 10** and **11**). The growing systems were 6-feet high (183 cm), each containing 16 42-inch (107 cm) long tubes, where six lettuce plants can grow using NFT, resulting in 96

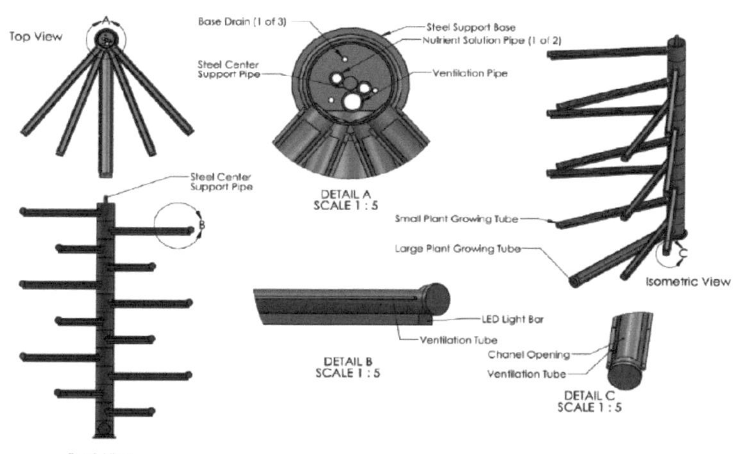

Figure 10.
The hydroponic growing tower system for the research greenhouse (left) and growing system in the CING (right).

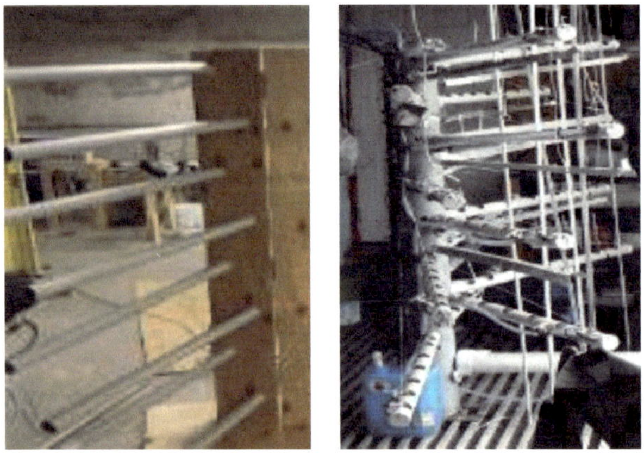

Figure 11.
Growing system prototype design described previously [6].

lettuce plants total per system. Tube diameters were 2 inches (5 cm) in diameter and lettuce heads were held in 2-inch (5 cm) net pots (**Figure 11**).

2.6.2 Flow in hydroponic systems

Each side of the growing systems has an independent pump. The nutrient solution is pumped by a magnetic drive submersible water pump (EcoPlus, Eco 396, US), delivering a flow of 1500 $L.h^{-1}$ (396 GPH), at a height of 2 m. A valve was used to control the flow in each tube, and a 1 $L.min^{-1}$ flow ensures a 3-mm level of nutrient solution in the 5 cm tubes [9]. Four NFT tubes per experiment were tested, to ensure 0.6–1 $L.min^{-1}$ per tube.

2.6.3 Electrical conductivity (EC)

EC was monitored with a handheld EC-meter (HM Digital Meters COM-80 Electrical Conductivity and Total Dissolved Solids Hydro Tester, Seoul, Korea). The EC was kept between 115–125 mS/m ($\pm$ 2.5 mS/m) above the greenhouse's irrigation water EC. The EC was adjusted by adding greenhouse irrigation water or concentrated nutrient solution [10]. pH

The pH of both nutrient solutions was maintained between 5.50 to 7.00 ($\pm$0.01). It was monitored with a handheld pH-meter (Dr. Meter PH100, China). Phosphoric acid (19.7% w/w) was used to lower pH to the desired value.

2.6.4 Light

Electrical light in the CING unit was provided by an LED installation. This comprised 10 light strips installed underneath the NFT tubes and six vertically hung light strips. When the thermal curtain was open, natural light was made available. In the Fall trial, the thermal curtain was only open when solar radiation was over 12 W/m^2 [7]. The outside light was measured with a Solar Radiation Smart Sensor (ONSET, Massachusetts, US), with a range of 0 to 1280 $W/m^2 \pm 10$ W/m^2. Light intensity to activate the thermal curtain was measured with a TSL2561 luminosity sensor, measuring Lux.

The natural lighting in the research greenhouse was supplemented with a high-pressure sodium (HPS) lamp lighting system. To ensure good growth, combined lighting is approximately 17 $mol/m^2/day$. The targeted instantaneous light intensity, measured with the LI-250A Quantum Radiometer Photometer, was estimated at

197 ± 1 μmol/m^2/sec. However, we expected that lighting would sometimes be lower than this targeted value, and the lowest light intensity value was estimated at 50 ± 1 μmol.m^{-2}.sec^{-1}. Lightmapping of the system was made to determine the amount of light achievable in both systems (Appendix A Tables A-5 to A-13) [10].

2.6.5 Temperature and relative humidity

The internal CING temperature set point was 24 °C during the day and 19 °C during the night time. This temperature was maintained using an electric auxiliary heater connected to an electrical thermostat (LUX Win100, Philadelphia, Pennsylvania). For the fall and winter trials. Auxiliary electrical heating was necessary and almost constant.

The internal temperature in the CING was monitored with a 12-Bit Temperature/Relative Humidity sensor (± 0.2 °C from 0° to 50 °C; ± 2.5% from 10% to 90%) compatible with the Hobo data logger (ONSET, Massachusetts, US). Humidity levels were not controlled.

The heating, ventilation and air-conditioning (HVAC) system were not functional for the test trials. However, exhaust fans were set on a thermostat, pulling fresh air into the CING, reducing temperature and relative humidity. A 9-inch 1100 CFM and a 16-inch 1435 CFM exhaust fan (Hessaire, Phoenix, Arizona, US) were mounted on the side wall, set on an electrical thermostat LUX Win100, Philadelphia, Pennsylvania) to cool the CING at 27 °C.

2.6.6 Crops

Romaine lettuce *(Lactuca sativa)* was cultivated for the first three trials (Spring 2018, Summer 2018 and Fall 2018), and Boston lettuce (*L. sativa)* was grown in Winter 2019 due to lack of available seeds.

2.7 Parameters

2.7.1 Light mapping

Lightmapping of the systems was made using a handheld Li-Cor Li-250A light sensor (LI-COR Biosciences, NE, US). To get the daily light integral (DLI) (mol. m^{-2}.d^{-1}), the photosynthetically active radiation (PAR) obtained at the brightest moment in the day was deducted from the PAR provided by the supplemental lights provided (PAR measurement after sundown), in the greenhouse and in the CING. PAR from the supplemental HPS lights in the greenhouse was 56.69 μmoles.m^{-2}.s^{-1} and PAR from the supplemental LED lights in the CING was 37.58 μmoles.m^{-2}.s^{-1}. Assuming that a quadratic function represents PAR versus the time of day for the

	Vermicompost Nutrient Solution				Hoagland Nutrient Solution			
Trial	**pH**	**EC (ms/m)**	**Temp. (°C)**	**Vol.(L)**	**pH**	**EC (ms/m)**	**Temp. (°C)**	**Vol. (L)**
1	9.1	129.9	31.7	13.8	7.9	160.2	30.3	12.2
2	6.4	140.8	26.4	15.0	6.5	146.8	26.4	15.5
3	6.9	109.5	22.6	12.9	6.6	118.4	21.9	12.4
4	5.1	146.7	24.0	14.5	4.9	84.5	23.1	11.3

Table 3.
Averages of monitored nutrient solution parameters for all trials (trial 1, 2, 3 and 4 respectively correspond to spring 2018, summer 2018, fall 2018 and winter 2019) in the research greenhouse.

	Vermicompost Nutrient Solution				Hoagland Nutrient Solution			
Trial	pH	EC (ms/m)	Temp. (°C)	Vol. (L)	pH	EC (ms/m)	Temp. (°C)	Vol. (L)
1	8.9	117.2	20.0	14.9	8.0	119.5	19.5	15.1
2	6.4	128.5	26.3	22.0	6.3	132.3	26.0	23.5
3	6.9	68.2	10.7	10.3	6.6	128.2	10.2	18.7
4	7.4	123.2	19.6	16.3	7.3	114.5	19.3	14.1

Table 4.
Averages of monitored nutrient solution parameters for all trials (trial 1, 2, 3 and 4 respectively correspond to spring 2018, summer 2018, fall 2018 and winter 2019) in the CING.

length of the specified day, with the measured PAR value at its highest value during daytime, it was possible to evaluate the maximum daily light integral from the Sunlight for a specific trial. By adding the DLI from the sun with the DLI of the supplemental light, a total maximum DLI was obtained.

For the Summer trial, PAR was measured on June 19th, 2018 under clear skies, assuming a 16-h day and 8-h night during the entirety of this trial. DLI in the greenhouse was evaluated at 29.4 mol/m^2/d and DLI in the CING was evaluated at 20.9 mol.m^{-2}.d^{-1}. For the Fall trial, PAR was measured on December 20th, 2018 under clear skies, assuming a day length of 8 h 50 min during this trial. DLI in the Fall in the greenhouse was evaluated at 5.1 mol.m^{-2}.d^{-1} and 7.61 mol.m^{-2}.d^{-1} in the CING. For the Winter trial, PAR was measured on March 19th, 2019 under clear skies, with an average daytime of 12 h, assuming the same PAR from supplemental lighting in the greenhouse and the CING from previous experiments. DLI in Winter in the greenhouse was evaluated at 18.0 mol.m^{-2}.d^{-1} and in the CING was evaluated at 9.3 mol.m^{-2}.d^{-1}. PAR mapping of the systems is available in Appendix A.

2.7.2 Monitoring of systems

The EC, pH, temperature and volume of the nutrient solutions for both systems were measured manually. Full monitoring data is available in the appendices and mean values for each trial are available in **Tables 3** and **4**.

2.8 Data analysis

Independent samples t-tests were performed using Excel to assess the statistical difference of the yields of fresh and dry masses of lettuce obtained in between growing environment for each trial.

3. Results

Season test Run	**Spring**				**Summer**			
Growth environment	**GH**		**CING**		**GH**		**CING**	
Treatment	V	H	V	H	V	H	V	H
Average fresh mass of lettuce (g)	0.82	33.63	0.64	4.60	4.81	53.25	1.86	7.41
S.E.	0.11	5.05	0.14	1.33	0.16	4.75	0.27	0.70
Season test Run	**Fall**				**Winter**			

Season test Run	**Spring**				**Summer**			
Growth environment	**GH**		**CING**		**GH**		**CING**	
Growth environment	GH		CING		GH		CING	
Treatment	V	H	V	H	V	H	V	H
Average fresh mass of lettuce (g)	2.51	17.54	0.99	0.97	4.38	23.40	2.07	16.79
S.E.	0.17	2.15	0.06	0.08	0.34	2.15	0.21	2.70

Table 5.
Average fresh mass with standard error (S.E.) for all treatments, greenhouse (GH) and CING, with Vermicompost (V) and Hoagland (H) nutrient solutions at harvest.

4. Discussion

4.1 Summary of results

Plants grown in the research greenhouse with the Hoagland nutrient solution had the highest fresh and dry mass for all tests (**Figure 12**). Of all the CING trials, the fresh and dry mass of lettuce grown in the CING with the Hoagland nutrient solution during the Winter trial was the highest (**Figure 13**). The Vermicompost nutrient solution had lower fresh and dry mass compared to the Hoagland in a common growing environment.

4.2 Environmental and growing parameters differences

Because of the climate difference between trials, the growth environment differed greatly in the CING. The lighting cycle for the Spring trial was 12 h day: 12 h night, the thermal curtain was active and roof panels were closed. In addition, pH was not controlled for this trial. The lighting cycle for the Summer trial was 12 h day: 12 h night, the thermal curtain was active and only one roof panel was open (**Figure 14**). The lighting cycle for the Fall trial was 16 h day: 8 h night, the thermal curtain was active and only one roof panel was open. The lighting cycle for the Winter trial was 24 h day 0 h night, the thermal curtain was not active and only one roof panel was open.

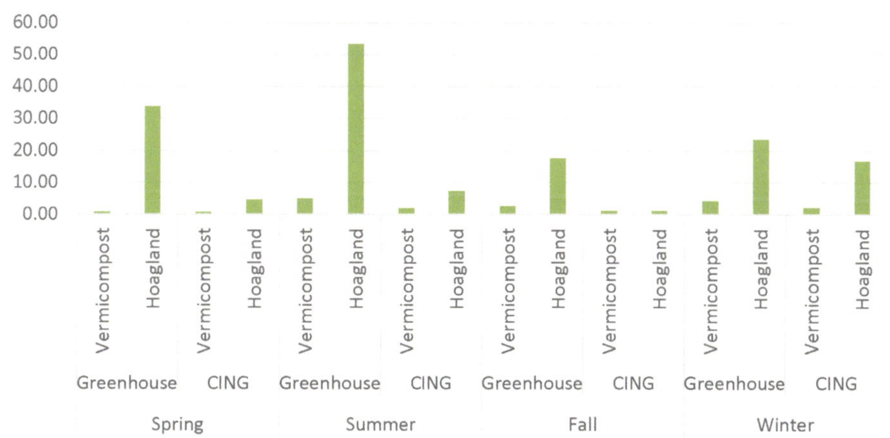

Figure 12.
Average fresh mass (g) of lettuce for all treatments at harvest.

Figure 13.
Lettuce grown in the CING before harvest, Winter 2019.

Figure 14.
Inside the CING, on top left is an opened roof panel, Summer 2018.

During the Spring trial, the pH in the vermicompost nutrient solution was over 8.5, pH was not controlled during the Spring trial and this may have limited nutrient availability and uptake.

During the Spring, Summer and Fall trials, plants in the CING grew very little when compared to plants grown in the greenhouse. During the Summer trial, the average temperature was slightly higher (25.4 °C) than the suggested temperature for lettuce growth (25 °C), and in the Fall the average temperature was 11 °C, which is lower than the recommended minimum (15 °C) for lettuce growth. Relative humidity for all trials ranged between of 50 to 70%, which is recommended for lettuce cultivation [10]. The Hoagland nutrient solution for the Winter trial was added at the beginning of the trial but not during; this explains the lower EC observed in the greenhouse for the Winter trial.

4.3 Cold weather trials

The Fall and Winter trials were the first cold climate trials undertaken in the CING unit. The comparison of the average conditions in the CING during both trial is available in the next table (**Table 6**).

For the Fall trial, the thermal curtain was set to open and close according to outdoor solar radiation. For the Winter trial, the thermal curtain remained closed, to help reduce thermal heat losses.

The curtain has an 80% shading level in diffused light PAR. The 20% of diffused light combined with the light from one opened roof panel, the constant supplemental lighting and the longer days allowed for greater DLI in the Winter Trial than the Fall trial. The average inside temperature in Fall was below the 15 °C recommended temperature for lettuce production [10]. This environmental difference explains the major difference in crop yield from the two cold conditions tests.

4.3.1 Thermal curtain

The thermal curtain usage changed the internal conditions of the CING. By comparing a set of days during both trials with similar outdoor temperature changes and environmental conditions, it is possible to better assess the impact of the thermal curtain. From December 10th to 12th 2018, the average outdoor and indoor temperatures were respectively, −7.6 °C and 12.3 °C. From March 4th to 6th 2019, the average outdoor and indoor temperatures were respectively, −8.2 °C and 7.5 °C.

Considering the thermal properties of the polycarbonate sheet, the thermal curtain and the insulating layer of air kept in between the thermal curtain and the polycarbonate sheet, with a temperature gradient of 15 °C from the inside and the outside of the CING the thermal heat loss from the window would be 17 Watts with the curtain closed, and 282 Watts with the curtain open. See the full heat transfer rate calculation in Appendix A.

Using the thermal curtain, the solar heat gain (SHG) to the CING was reduced, proportionally to the sunlight blocked, 80% [11]. This difference in SHG can be linked to the more stable temperature during the day, noticeable in **Figure 15** during the Fall trial cold days testing. However, during the Winter trial, with the thermal curtain constantly closed, the inside temperature was more dependent on the outside temperature as observed in **Figure 16** for a 3 days comparison with similar average temperatures.

This trend can be observed when comparing the relationship between the indoor and outdoor temperatures, during the 3 days comparison in **Figures 17** and **18** and the whole experiment data in **Figures 19** and **20**. Whereas the R^2 = 0.0656 for the

Trial	Average Outside Temperature (°C)	Average Inside Temperature (°C)	Approximate DLI ($mol.m^{-2}.d^{-1}$)	Average Fresh Mass (g)
Fall 2018	−3.9	11.0	7.6	0.97
Winter 2019	−2.4	14.8	9.3	16.79

Table 6.
*Summary of **Table 3**, **Table 4** and **Table 5** for cold condition trials of the CING.*

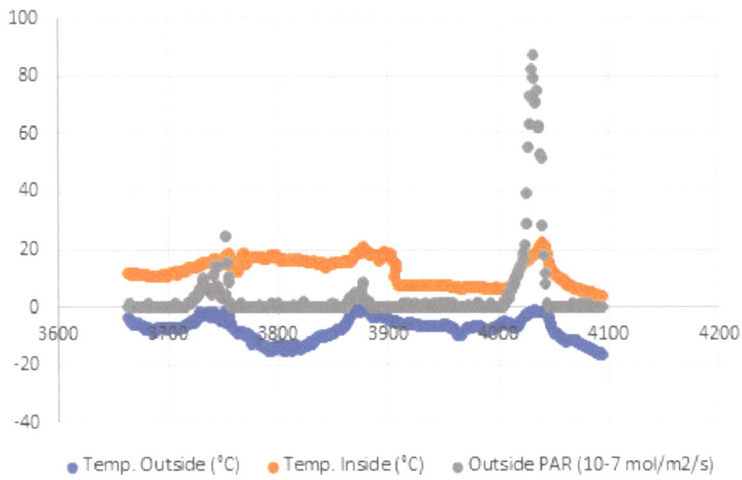

Figure 15.
Outside temperature, inside temperature and outside PAR of the CING, December 10th to December 12th 2018.

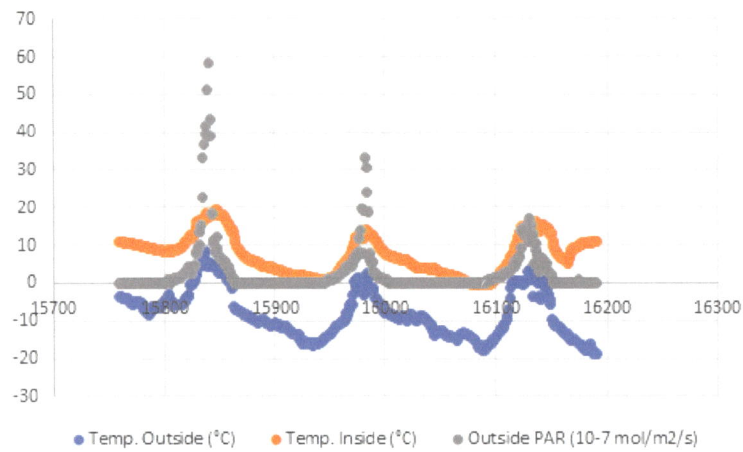

Figure 16.
Outside temperature, inside temperature and outside PAR of the CING, march 4th to march 6th 2019.

Fall trial and R^2 = 0.702 for the Winter trial during the 3 days comparison and R^2 = 0.3114 for the Fall trial and R^2 = 0.5741 for the Winter trial during the full trials.

4.3.2 Energy usage

Considering that the average cold and warm weather maximum energy requirements of the CING are approximately 21.7 $kWh.m^{-2}$, the maximum yearly energy use of the CING would be 7920 $kWh.m^{-2}$. This is still considerably higher than the modified shipping container described by The University of Arizona and higher than the 711.91 $kWh.m^{-2}$ average for 164 greenhouses occupying a total of 16444 m^2 operated by Cornell University's Agricultural Experiment Station in New York [6].

The use of the thermal curtain showed an effect on inside temperature, but the extra sunlight SHG did not provide enough light and heat to achieve growing parameters during the Fall trial. The use of electrical lights and heating however provided enough light and heat to achieve growing parameters during the Winter trial.

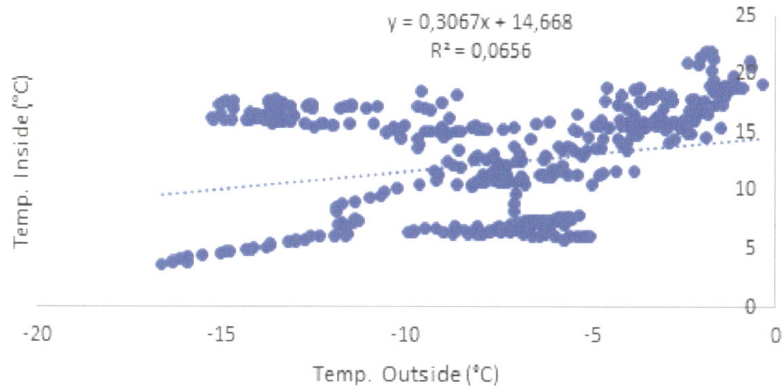

Figure 17.
Temperature inside vs. temperature outside of CING, fall trial, December 10th to 12th 2018.

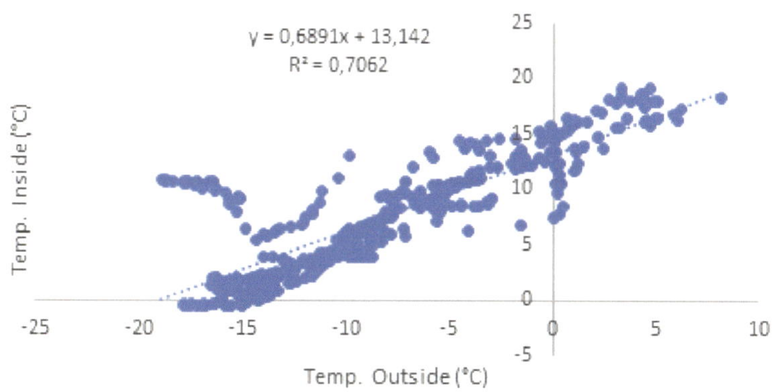

Figure 18.
Temperature inside vs. temperature outside of CING, winter trial, march 4th to 6th 2019.

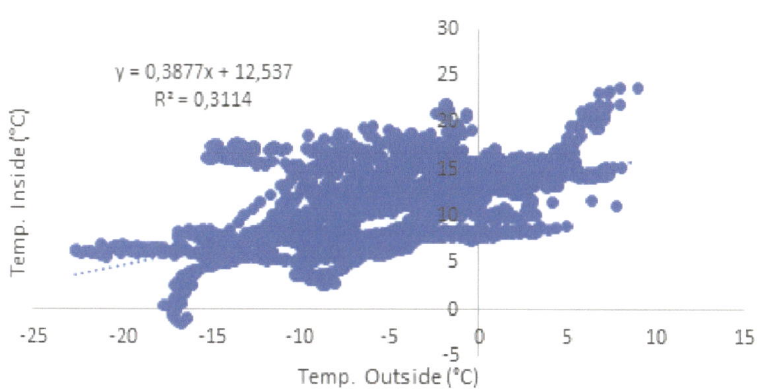

Figure 19.
Temperature inside vs. temperature outside of CING, fall trial, December 1st to December 22nd 2018.

Heating was almost constant in cold conditions, with an average indoor temperature for the Winter trial of 14.8 °C. Heating was the most energy-intensive parameter of the CING, representing 62% of the maximum daily energy requirement, but the achieved temperature was still lower than the recommended temperature for lettuce growth [10].

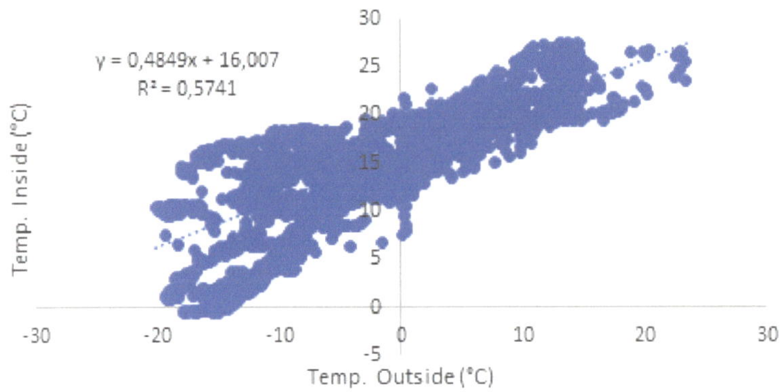

Figure 20.
Temperature inside vs. temperature outside of CING, winter trial, march 1st to march 23rd 2019.

4.3.3 Other considerations

The CING structure was strong enough to withstand the weight of snow accumulation.

Interestingly, we observed that the highest lettuce yield for the CING-grown plants was during the Winter trial. This demonstrates the potential of winter growth within the CING.

The vermicompost-based nutrient solution has seen an improvement from the beginning of the experiments but the nutrient profile is not yet complete and provides lower lettuce yields than the Hoagland nutrient solution.

4.4 Feasibility of the CING

Inspired by container farming, the CING was designed to operate in a cold and warm climate, exemplified by the short growing season in northern Canada. The environmental conditions surrounding the CING had a major impact on its interior environment, but the ability to insulate the CING unit using a thermal curtain helped manage heat and keep stable growing conditions.

If CF can successfully allow for food crop growth in a cold climate as demonstrated by these CING trials, the prototype cannot yet be considered viable as heating demands are too high and environmental control is not adequate. However, the use of natural light has made it possible to cultivate plants in this growing environment with minimal supplemental lighting. The main issue with the CING is its capacity to keep a desired internal temperature under outdoor cold conditions. The opening of the thermal curtain did increase light intensity and allowed for a higher solar heat gain. Performance of the CING in terms of biomass production was higher when the thermal curtain remained closed during the Winter, but this result is mainly caused by the average inside temperature and DLI to be higher during this trial.

5. Conclusion

The CING unit was able to successfully grow lettuce plants in a cold climate during the Winter trial but energy demands were still very high because of heating.

The dry mass of lettuce grown in the winter achieved 72% of the average fresh mass of lettuce grown at the same time in the greenhouse. In addition, the lettuce grown in the CING during the winter had the highest fresh and dry mass when compared to the other trials in the CING unit when using Hoagland nutrient solution. The vermicompost nutrient solution allowed for lettuce growth but at a much lower yield for all trials likely due to nitrogen deficiency. Continuous supplemental LED light provided the best results for lettuce growth in the CING. The thermal curtain opening according to an outdoor solar radiation threshold did allow for more light and heat in the CING unit, reducing the correlation of inside and outside temperature, under cold outdoor conditions.

5.1 Recommendations

The combination of natural and supplemental light in CF has the potential to reduce energy needs linked to lighting. However, heat loss analyses must be made to evaluate the energy efficiency of a single transparent wall or part of a single transparent wall of a container farm in a northern Canada climate.

Secondly, trials performed in the CING only used a small part of the growing space. To decrease the energy needs per growth surface another hydroponic configuration could be used. Container farms often use stacked shallow water cultures to grow leafy greens, which allows the highest density of crop production. Considering the full growing area of the CING represents half of a 40-foot shipping container or 14.4 m^2, 75% of this the growing area or 10.8 m^2 could be used for plant growth, thus reducing energy requirements per square meter of production. More lighting and air exchange would be needed to use all the growing areas, and heating energy requirements might be reduced by the addition of supplemental lighting. Modifying the CING for better space usage could reduce energy demands per unit of crops produced.

Thirdly, a recommended modification to the CING unit would be better environmental control, with a functional HVAC system; to increase the temperature and humidity control of the CING. Plus a larger thermal mass of the northern wall of the CING; to reduce the heating requirements by increasing the passive heating of the CING [12]. A complete heat exchange simulation of the CING would be necessary to compare its performance as a northern growing unit.

Appendix A

A.1 Monitoring of systems

	GREENHOUSE											CING										
	Vermicompost Nutrient Solution					Hoagland Nutrient Solution						Vermicompost Nutrient Solution					Hoagland Nutrient Solution					
DATE	pH	EC (ms/m)	T_solution (°C)	Level (inch)	Volume (L)	pH	EC (ms/m)	T_solution (°C)	Level (inch)	Volume (L)	T_amb (°C)	pH	EC (ms/m)	T_solution (°C)	Level (inch)	Volume (L)	pH	EC (ms/m)	T_solution (°C)	Level (inch)	Volume (L)	T_amb (°C)
2018-05-08	8.4	110.0	36.2			7.5	150.0	36.2			28.5	8.2	87.7	21.1			7.5	131.0	21.0			27.0
2018-05-09	8.4	106.7	34.6			7.3	154.5	33.0			31.0	8.3	89.2	22.3			7.5	134.2	21.4			25.0
2018-05-11	9.0	100.0	32.0			7.7	160.0	32.0			34.0	8.7	81.0	13.0			7.9	126.0	13.0			13.0
2018-05-14	9.2	100.0	32.0			7.7	160.0	32.0			34.0	8.7	81.0	13.0			7.9	126.0	13.0			13.0
2018-05-15	9.7	123.0	26.0	3.3	13.2	7.7	160.0	26.0	4.0	16.3	28.0	9.0	109.0	18.0	4.0	16.3	8.1	124.0	18.0	4.0	16.3	20.0
2018-05-16	9.2	134.0	35.0	3.0	12.2	7.9	168.0	34.0	3.5	14.2	35.0	8.9	112.0	14.0	3.8	15.4	8.1	126.0	14.0	3.8	15.4	22.0
2018-05-17	9.1	141.0	27.0	2.9	11.8	7.9	183.0	27.0	3.4	13.8	28.0	9.0	116.0	21.0	3.9	15.9	8.3	127.0	21.0	3.8	15.4	22.0
2018-05-18	9.3	134.0	33.0	3.5	14.2	8.2	172.0	32.0	3.5	14.2	27.0	9.1	126.0	24.0	3.6	14.6	8.4	123.0	23.5	4.3	17.3	28.0
2018-05-21	9.3	153.0	32.0	2.8	11.4	8.3	216.0	31.0	2.2	8.9	29.0	9.0	135.0	19.0	3.4	13.8	8.3	129.0	18.0	3.8	15.4	22.0
2018-05-22	9.2	131.0	28.0	3.7	15.0	8.3	133.0	26.0	4.0	16.3	28.0	9.1	135.0	24.0	3.5	14.2	7.7	100.0	22.0	5.0	20.3	20.0
2018-05-23	9.2	140.0	32.0	3.5	14.2	8.3	143.0	31.0	3.5	14.2	27.0	9.1	132.0	20.0	3.5	14.2	8.1	102.0	20.0	4.0	16.3	22.0
2018-05-24		147.0	32.0	3.5	14.2		162.0	30.0	2.5	10.2	28.0		142.0	24.0	3.3	13.2		105.0	24.0	2.8	11.2	24.0
2018-05-25		157.0	32.0	3.2	13.0	7.9	211.0	26.0	1.5	6.1	26.0	9.2	145.0	26.0	3.4	13.8	8.3	110.0	24.0	1.3	5.3	26.0
2018-05-28	9.0	131.0		4.0	16.3	7.7	110.0		3.0	12.2			131.0		4.3	17.3		115.0		4.5	18.3	24.0
2018-05-29		141.0	32.0	4.0	16.3		121.0	28.0	2.0	8.1	23.0		136.0					115.0				
AVERAGE	9.1	129.9	31.7	3.4	13.8	7.9	160.2	30.3	3.0	12.2	29.0	8.9	117.2	20.0	3.7	14.9	8.0	119.5	19.5	3.7	15.1	22.0

Table A-1.
Monitoring of pH, EC, temperature and volume of nutrient solution for the Spring trial.

	GREENHOUSE											CING										
	Vermicompost Nutrient Solution					Hoagland Nutrient Solution						Vermicompost Nutrient Solution					Hoagland Nutrient Solution					
DATE	pH	EC (ms/m)	T_solution (°C)	Level (inch)	Volume (L)	pH	EC (ms/m)	T_solution (°C)	Level (inch)	Volume (L)	T_amb (°C)	pH	EC (ms/m)	T_solution (°C)	Level (inch)	Volume (L)	pH	EC (ms/m)	T_solution (°C)	Level (inch)	Volume (L)	T_amb (°C)
2018-06-12	7.1	176.0	26.8	2.5	10.2	7.3	191.0	27.3	2.0	8.1		6.8	127.0	24.0	5.5	22.4	6.7	138.0	24.0	5.3	21.3	
2018-06-13	6.9	144.0	27.1	4.3	17.3	7.1	158.0	27.1	4.6	18.7		6.8	131.0	27.0	5.0	20.3	6.8	141.0	27.0	4.9	19.9	
2018-06-19												6.2	135.0				6.4	134.0		5.3	21.3	
2018-06-21	6.1	131.0		4.0	16.3	6.1	135.0		4.0	16.3		6.1	134.0		5.5	22.4	5.9	136.0		6.0	24.4	
2018-06-26	6.2	142.0	25.2	3.8	15.2	6.2	141.0	24.9	4.0	16.3		6.3	133.0	27.1	4.8	19.3	6.1	136.0	26.3	5.5	22.4	
2018-06-28	5.9	111.0		4.0	16.3	5.9	109.0		4.5	18.3		6.0	111.0	26.9	6.3	25.4	5.9	109.0	26.5	7.8	31.5	
AVERAGE	6.4	140.8	26.4	3.7	15.0	6.5	146.8	26.4	3.8	15.5		6.4	128.5	26.3	5.4	22.0	6.3	132.3	26.0	5.8	23.5	

Table A-2.
Monitoring of pH, EC, temperature and volume of nutrient solution for the Summer trial.

	Vermicompost Nutrient Solution					Hoagland Nutrient Solution						Vermicompost Nutrient Solution					Hoagland Nutrient Solution					
DATE	pH	EC (ms/m)	T_solution (°C)	Level (inch)	Volume (L)	pH	EC (ms/m)	T_solution (°C)	Level (inch)	Volume (L)	T_amb (°C)	pH	EC (ms/m)	T_solution (°C)	Level (inch)	Volume (L)	pH	EC (ms/m)	T_solution (°C)	Level (inch)	Volume (L)	T_amb (°C)
2018-12-01	6.9	95.0		3.5	14.2	6.6	123.0		4.0	16.3		6.8	80.0		2.0	8.1	6.5	130.0		2.5	10.2	
2018-12-04	8.1	105.0	22.8	3.0	12.2	6.9	133.0	22.7	3.5	14.2	22.0	7.4	47.0	11.9	4.5	18.3	7.1	63.0	11.6	6.0	24.4	13.9
2018-12-04	7.2	110.0	22.4	3.0	12.2					0.0						0.0	6.6	129.0	11.7	6.5	26.4	13.9
2018-12-05	7.0	117.0	22.2	3.0	12.2	7.0	128.0	22.9	3.0	12.2		7.0	58.0	8.5	5.0	20.3	7.0	130.0	8.4	6.0	24.4	10.0
2018-12-10	7.0	107.0	23.1	3.0	12.2	6.7	120.0	19.9	3.0	12.2		6.9	67.0	9.0	4.5	18.3	6.9	134.0	9.1	5.0	20.3	13.3
2018-12-11	7.5	117.0	21.6	2.5	10.2	7.2	130.0	21.9	2.5	10.2	20.2	7.7	73.0	7.0	4.8	19.3	7.0	137.0	7.4	6.0	24.4	8.9
2018-12-13	6.5	122.4		3.0	12.2	6.3	93.5		3.0	12.2		6.5	57.8	11.1		0.0	6.1	105.3	8.8	5.5	22.4	16.1
	6.5	125.0		3.0	12.2	6.0	125.8		4.0	16.3		6.6	71.8			0.0	6.0	136.4			0.0	
2018-12-17	6.6	93.9	21.3	4.0	16.3	6.5	105.5	19.2	4.0	16.3	18.8	6.7	74.1	13.3	4.0	16.3	6.2	151.6	11.9	5.0	20.3	15.6
2018-12-18	6.9	95.0	23.9	4.0	16.3	6.5	112.0	23.8	4.0	16.3						0.0				4.0	16.3	
2018-12-21	6.5	116.9	23.2	3.0	12.2	6.5	113.4	22.8	2.5	10.2		6.9	84.9	13.8	3.0	12.2	6.3	165.3	12.4	4.0	16.3	15.0
AVERAGE	6.9	109.5	22.6	3.2	12.9	6.6	118.4	21.9	3.4	12.4	20.3	6.9	68.2	10.7	4.0	10.3	6.6	128.2	10.2	5.1	18.7	13.3

Table A-3.
Monitoring of pH, EC, temperature and volume of nutrient solution for the Fall trial.

DOI: http://dx.doi.org/10.5772/intechopen.96214

	GREENHOUSE													CING												
	Vermicompost Nutrient Solution						Hoagland Nutrient Solution							Vermicompost Nutrient Solution						Hoagland Nutrient Solution						
DATE	pH	Tds (ppm)	EC (ms/m)	T_solution (°C)	Level (inch)	Volume (L)	pH	Tds (ppm)	EC (ms/m)	T_solution (°C)	Level (inch)	Volume (L)	T_amb (°C)	pH	Tds (ppm)	EC (ms/m)	T_solution (°C)	Level (inch)	Volume (L)	pH	Tds (ppm)	EC (ms/m)	T_solution (°C)	Level (inch)	Volume (L)	T_amb (°C)
2019-03-02	5.6	737.0	140.0	22.7	3.0	12.2	5.6	676.0	128.4	22.1	3.0	12.2		7.6	705.0	134.0	15.0	3.0	12.2	7.2	602.0	114.4	14.6	3.0	12.2	
2019-03-05	4.2	783.0	148.8	26.0	2.8	11.2	2.5	805.0	153.0	25.2	2.0	8.1				0.0										
	4.3	674.0	128.1		3.5	14.2	4.3	680.0	129.2		2.5	10.2		6.1	567.0	107.7	5.3	2.5	10.2	6.6	546.0	103.7	6.6	2.8	11.2	11.1
2019-03-06	4.3	599.0	113.8	23.7	5.0	20.3	4.7	515.0	97.9	24.2	3.5	14.2		7.0	697.0	132.4	20.3	5.0	20.3	7.1	575.0	109.3	20.0	5.0	20.3	
2019-03-11	5.3	671.0	127.5	21.4	4.8	19.5	4.8	362.0	68.8	18.6	3.5	14.2		7.3	910.0	172.9	25.0	2.3	9.1	7.3	792.0	150.5	23.5	2.5	10.2	
														7.8	649.0	123.3		4.5	18.3	7.4	662.0	125.8		3.0	12.2	
2019-03-14	5.1	923.0	175.4	25.8	2.3	9.1	4.9	348.0	66.1	24.9	2.0	8.1		7.6	776.0	147.4	24.2	3.5	14.2	7.4	780.0	148.2	23.9	2.0	8.1	
	5.0	770.0	146.3	22.1	4.5	18.3	4.8	269.0	51.1	23.4	4.0	16.3		7.6	705.0	134.0	21.5	5.3	21.3	7.5	522.0	99.2	20.9	3.0	12.2	
2019-03-19	4.5	889.0	168.9	24.3	2.8	11.2	5.0	168.0	31.9	21.3	2.5	10.2		7.5	961.0	182.6	24.8	4.0	16.3	7.5	815.0	154.9	24.0	2.0	8.1	
														7.5	660.0	125.4	21.0	6.0	24.4	7.5	363.0	69.0	18.0	6.5	26.4	
2019-03-25	7.8	901.0	171.2	26.0	3.5	14.2	8.0	180.0	34.2	25.2	2.0	8.1		8.0	500.0	95.0	19.5	4.0	16.3	7.8	370.0	70.3	22.0	5.0	20.3	
AVERAGE	5.1	771.9	146.7	24.0	3.6	14.5	4.9	444.8	84.5	23.1	2.8	11.3		7.4	713.0	123.2	19.6	4.0	16.3	7.3	602.7	114.5	19.3	3.5	14.1	11.1

Table A-4.
Monitoring of pH, EC, temperature and volume of nutrient solution for the Winter trial.

A.2 Temperature monitoring of the CING

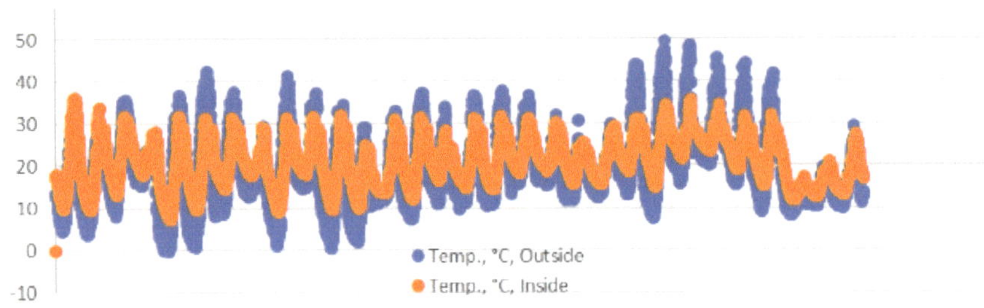

Figure A.1.
Temperature monitoring outside and inside the CING, Spring trial, corresponding averages : 19.3°C and 21.2°C.

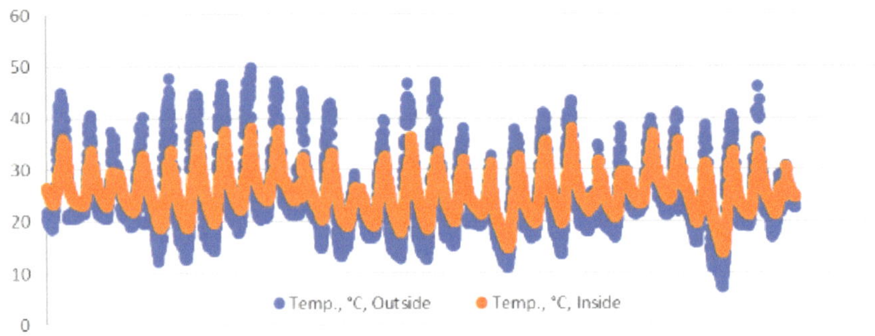

Figure A.2.
Temperature monitoring outside and inside the CING, Summer trial, corresponding averages: 24.7°C and 25.4°C.

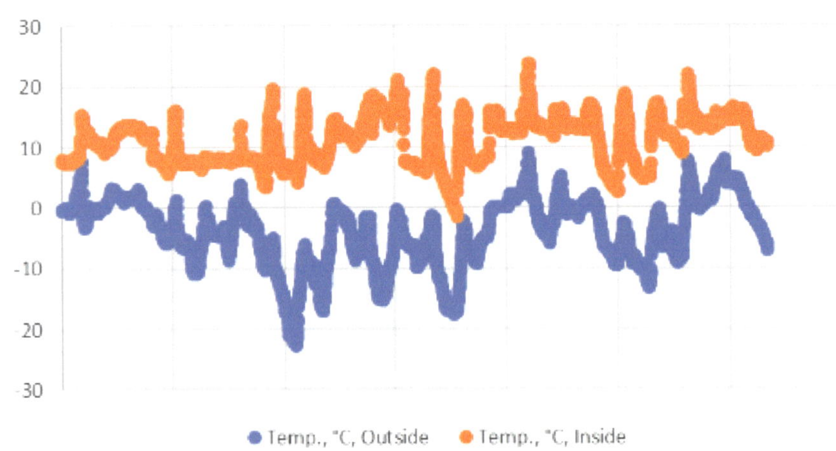

Figure A.3.
Temperature monitoring outside and inside the CING, Fall trial, corresponding averages: -3.4°C and 11.0°C.

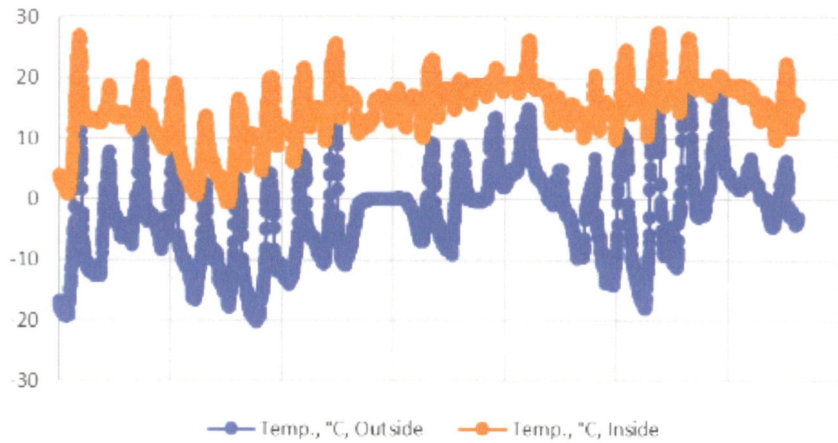

Figure A.4.
Temperature monitoring outside and inside the CING, Winter trial, corresponding averages: -2.4°C and 14.8°C.

A.3 Humidity monitoring of the CING

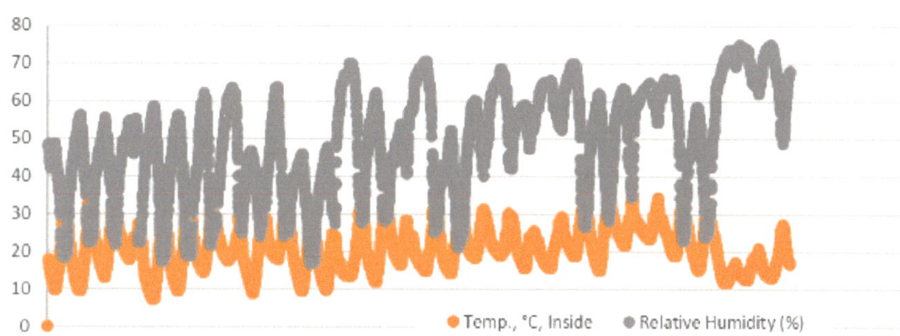

Figure A.5.
Humidity and temperature monitoring inside the CING, Spring trial, average relative humidity: 49.2 %.

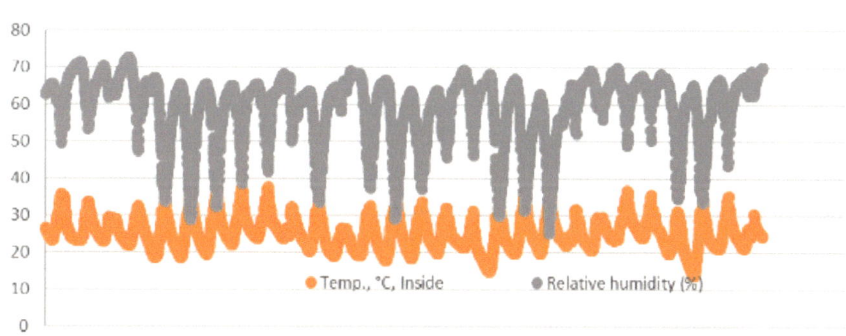

Figure A.6.
Humidity and temperature monitoring inside the CING, Summer trial, average relative humidity: 59.1 %.

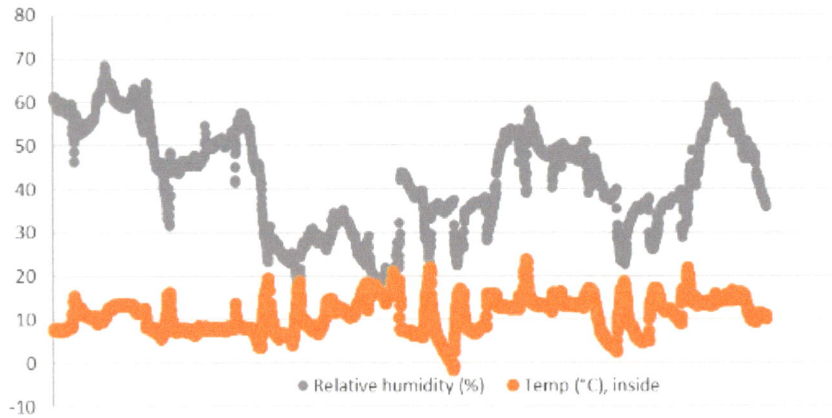

Figure A.7.
Humidity and temperature monitoring inside the CING, Winter trial, average relative humidity: 35.1 %.

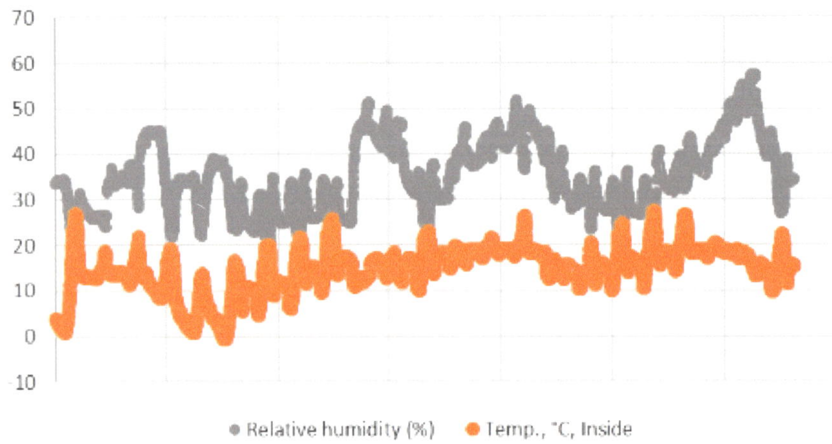

Figure A.8.
Representation of the thermal resistance of the different layers of the CING window (Bergman, Lavine, Incropera, & Dewitt, 2011).

A.4 Light mapping of systems

Experiment	**Greenhouse**	
Date	2018-06-19	
Time	12:20	
Weather	Very sunny	
	PAR	µmoles/m²/s
Row	Left	Right
1	322	962
2	669	681
3	709	1077
4	937	699

Experiment	Greenhouse	
Average	659.25	854.75
Average PAR	757	

Table A-5.
Light mapping, Summer trial.

Experiment	CING	
Date	2018-06-19	
Time	12:20	
Weather	Very sunny	
	PAR	µmoles/m^2/s
	Left Row	Right rows
1	179	538
2	525	511
3	434	194
4	599	806
Average	434.25	512.25
Average PAR	473.25	

Table A-6.
Light mapping, Summer trial.

Experiment	CING	
Date	2018-06-19	
Time	14:20	
Weather	Very sunny	
	PAR	µmoles/m^2/s
	Left Row	Right row
1	276.5	259.3
2	523.2	356.7
3	802.9	531.7
4	832.6	781.1
Average	608.8	482.2
Average PAR	545.5	

Table A-7.
Light mapping, Summer trial.

Experiment	Greenhouse	
Date	2018-12-20	
Time	19:00	
Weather	Night	
	PAR	µmoles/m2/s
Row	Left	Right
1	51.6	29.14
2	43.68	43.65
3	65.87	51.87
4	86.31	81.37
Average	61.87	51.51
Average PAR	56.69	

Table A-8.
Light mapping, Fall trial, only supplemental light in the greenhouse.

Experiment	Greenhouse	
Date	2018-12-20	
Time	14:30	
Weather	Very sunny	
	PAR	µmoles/m2/s
Row	Left	Right
1	76.14	76.2
2	66.27	73.3
3	88.2	98.53
4	114.92	112.23
Average	86.38	90.07
Average PAR	88.22	

Table A-9.
Light mapping, Fall trial only.

Experiment	CING	
Date	2018-12-20	
Time	19:00	
Weather	Night	
	PAR	µmoles/m2/s
Row	Left	Right
1	48.09	63.52
2	57.23	59.76

Experiment	**CING**	
3	12.57	20.52
4	20	18.94
Average	34.47	40.69
Average PAR	37.58	

Table A-10.
Light mapping, Fall trial, supplemental light in the CING.

Experiment	**CING**	
Date	2018-12-20	
Time	15:00	
Weather	Very Sunny	
	PAR	µmoles/m2/s
Row	Left	Right
1	53.5	278
2	145.08	509
3	166.34	506.4
4	187.46	523.3
Average	138.10	454.18
Average PAR	296.14	

Table A-11.
Light mapping, Fall trial.

Experiment	**Greenhouse**	
Date	2019-03-19	
Time	13:00	
Weather	Clear sky	
	PAR	µmoles/m2/s
Row	Left	Right
1	348.10	685.70
2	598.00	536.90
3	498.50	580.60
4	638.90	670.20
Average	520.88	618.35
Average PAR	569.61	

Table A-12.
Light mapping, Greenhouse Winter trial.

Experiment	CING	
Date	2019-03-19	
Time	13:30	
Weather	Clear sky	
	PAR	µmoles/m2/s
Row	Left	Right
1	110.96	174.20
2	261.50	257.80
3	59.44	197.55
4	475.30	452.10
Average	226.80	270.41
Average PAR	248.61	

Table A-13.
Light mapping, CINGWinter trial.

A.5 Thermal curtain heat transfer rate calculation

Heat transfer rate calculation

$$q_x = \frac{T_{\infty,1} - T_{\infty,4}}{[(1/h_1 A) + (L_A/k_A A) + (L_B/k_B A) + (L_C/k_C A) + (1/h_4 A)]}$$

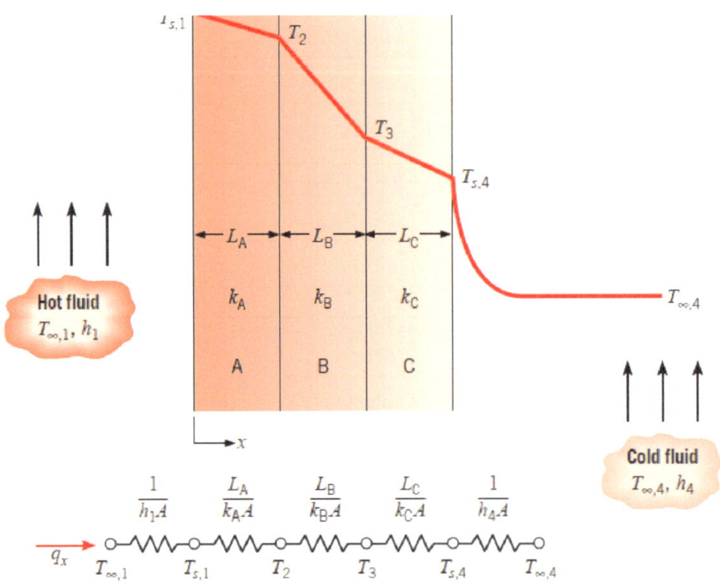

Figure A.9.
Humidity and temperature monitoring inside the CING, Fall trial, average relative humidity: 42.2 %.

Parameters	Value
Convective heat transfer coefficient of air inside CING, h_1 (W/(m^2.K)	20 (EngineeringToolBox, 2020)
Convective heat transfer coefficient of air outside CING h_4 (W/ m^2.K)	30 (EngineeringToolBox, 2020)
Thermal conductivity of thermal curtain, k_A (W/m.K)	0.104 (AZOMaterials, 2020) and (Ludvig Svensson, 2020)
Thermal conductivity of air layer, k_B (W/m.K)	25.3x10^{-3} (Bergman, Lavine, Incropera, & Dewitt, 2011)
Thermal conductivity of Twin-Wall polycarbonate Sheet, k_C (W/m.K)	37.86 (PALRAM, 2010)
Thickness of Curtain, L_A (m)	0.001
Thickness of air Layer, L_B (m)	0.15
Thickness of Twin Wall polyecarbonate sheet, L_C (m)	0.008
Area of Window (m^2)	7.27
Temperature gradient, $T_{\infty,1}$ -$T_{\infty,A}$ (K)	15.0

Table A-14.
Parameters for heat transfer rate calculations.

Heat transfer rate, qx (Watts) without curtain and stagnant air layer	282.4
Heat transfer rate, qx (Watts) with curtain and stagnant air layer	17.2

Table A-15.
Heat transfer rate calculation result.

Author details

David Leroux* and Mark Lefsrud
McGill University, Sainte-Anne-de-Bellevue, Canada

*Address all correspondence to: david.leroux3@mail.mcgill.ca

References

[1] Ramin Shamshiri, R., Kalantari, F., C. Ting, K., R. Thorp, K., A. Hameed, I., Weltzien, C., . . . Mojgan Shad, Z. (2018). Advances in greenhouse automation and controlled environment agriculture: A transition to plant factories and urban agriculture. International Journal of Agricultural and Biological Engineering, 11(1), 1–22.

[2] Touliatos, D., Dodd, I. C., & McAinsh, M. (2016). Vertical farming increases lettuce yield per unit area compared to conventional horizontal hydroponics. Food Energy Secur, 5(3), 184–191.

[3] Agrilyst. (2017). State of Indoor Farming.

[4] Benis, K., Reinhart, K., & Ferrão, P. (2017). Building-Integrated Agriculture (BIA) In Urban Contexts: Testing A Simulation-Based Decision Support Workflow. Paper presented at the International Building Performance Simulation Association, San Francisco.

[5] MIT. (2016). Leafy Green Machine Business Feasability Evaluation. Retrieved from Laboratory for Sustainable Business:

[6] Liu, X. (2014). Design of a Modified Shipping Container as Modular Unit for the Minimally Structured & Modular. (Master of Science). University of Arizona.

[7] Gaudet, P. (2017). Food Security in Northern Canada (FOOD SINC) Unit: Weather Data and Environment Control Analysis for the Determination of Automation System Parameters. Faculty of Agricultural and Environmental Sciences, McGill University.

[8] Fernandez, D. (Producer). (2009, February 2). The Hoaglands Solution for Hydroponic Cultivation. Science in Hydroponics.

[9] Lennard, W. A., & Leonard, B. V. (2006). A comparison of three different hydroponic sub-systems (gravel bed, floating and nutrient film technique) in an Aquaponic test system. Aquacult Int, 539–550.

[10] Brechner, M., & Both, A. J. (2013). Hydroponic Lettuce Handbook. Cornell University.

[11] Ludvig Svensson. (2020). TEMPA7567DFB CS Product Sheet.

[12] Beshada, E., Zhang, Q., & Boris, R. (2006). Winter performance of a solar energy greenhouse in southern Manitoba. Canadian Biosytems Engineering, 49, 5.1–5.8.

Chapter 4

Radiation Exchange at Greenhouse Tilted Surfaces under All-Sky Conditions

Erick K. Ronoh

Abstract

Greenhouses generally exhibit a greater degree of thermal radiation interaction with the surroundings than other buildings. A number of greenhouse thermal environment analyses have handled the thermal radiation exchange in different ways. Thermal radiation exchange at greenhouse surfaces is of great interest for energy balance. It dominates the heat transfer mechanisms especially between the cover material surface and the surrounding atmosphere. At these surfaces, the usual factors of interest are local temperatures and energy fluxes. The greenhouse surfaces are inclined and oriented in various ways and thus can influence the radiation exchange. The scope of this work is determination of the thermal radiation exchange models as well as effects of surface inclination and orientation on the radiation exchange between greenhouse surfaces and sky. Apart from the surface design and the thermal properties of the cover, the key meteorological parameters influencing longwave and shortwave radiation models were considered in detail. For the purpose of evaluating surface inclination and orientation effects, four identical thermal boxes were developed to simulate the roof and wall greenhouse surfaces. The surface temperatures and atmospheric parameters were noted under all-sky conditions (clear-sky and overcast). Differences in terms of surface-to-air temperature differences at the exposed roof and wall surfaces as influenced by surface inclination and orientation are discussed in this work. Overall, the findings of this work form a basis for decisions on greenhouse design improvements and climate control interventions in the horticultural industry.

Keywords: Radiation exchange, Greenhouses, Tilted surfaces, Roof, Wall, Sky

1. Introduction

Thermal radiation dominates the heat transfer mechanisms especially between the cover material surface and the surrounding atmosphere. The radiation heat transfer depends on the orientation of the surfaces relative to each other as well as their radiation properties and temperatures [1]. For a non-horizontal surface (e.g. roof and wall), the radiation exchange between the surface and the sky is weighted by a view factor. The view factor gives the fraction of the view from a base surface obstructed by a given other surface [2]. Generally, single-span greenhouses are oriented such that the length runs east–west. This orientation maximizes winter sunlight and heat gain in the greenhouse [3]. Gutter-connected greenhouses are

Figure 1.
Multi-span Venlo glass greenhouse in Hannover, Germany.

oriented with the length running north–south (**Figure 1**). According to Sanford [3], this ensures that the shadow cast by the gutters moves during the day. If the orientation is east–west in this case, the shadow of the gutter will move very little, resulting in less direct sunlight and thus slowing down the plant growth. Spatial irregularities of irradiance with east–west oriented greenhouses could often be a problem at all latitudes [4]. Generally, a specific orientation is suitable for a given purpose and location.

The precise determination of the radiation components is essential for a good estimate of the net radiation balance and, consequently, of the radiation and energy balances [5]. The radiation balance, the main source of energy available for the physical and biological processes, is the essential component of the energy balance at the surface. With the availability of hydro-meteorological data such as air temperature, relative humidity, and cloudiness, the longwave radiation can be estimated for any location and at any given time. The understanding of the factors which control the ascending and descending flows in the atmosphere is essential to improve the models used in the various environmental applications [6].

Internationally, a substantial emphasis is placed on a greenhouse orientation that maximizes light interception. At different surface inclinations and orientations, accurate radiation data and models are required for the longwave radiation exchange at representative conditions [7]. Hence, this chapter seeks to determine the influence of the glass-covered greenhouse surface inclination and orientation on the exterior longwave radiation exchange.

2. Thermal radiation exchange at greenhouse surfaces

Generally, there is greater thermal radiation interaction between greenhouses and the surroundings compared to other buildings. As a result, thermal radiation loss can particularly become the dominant mechanism of total heat loss especially at night. Thermal radiation is therefore a very important factor in determining the thermal environment inside a greenhouse. Simulation models help to address the challenges related to the high costs of directly measuring longwave radiation. The simulation models further allow the estimation of the thermal exchange on any building surfaces. It is evident from the models that neglecting to consider thermal radiation (shortwave and longwave) exchange in sufficient detail can lead to serious inaccuracies in the model predictions. For the energy balance under day-time conditions, the solar irradiance on greenhouse surfaces plays a very important role and should, therefore, be accounted for precisely. The solar radiation data is readily available from most weather stations particularly for horizontal surfaces and this, together with other parameters, can be utilized in calculating the total irradiance on tilted surfaces with an acceptable accuracy. Knowledge of the thermal radiation exchange is vitally important for numerous applications in agriculture requiring surface radiation and energy balance.

2.1 Longwave radiation exchange

Modeling of longwave radiation exchange between the outside surfaces and the sky requires the knowledge of the sky temperature [8]. The equivalent sky temperature T_{sky} has been estimated differently by various researchers. The common equations applied in the T_{sky} computation are empirical in nature and are related to the air temperature. Thus, they perform best for areas with a radiative climate similar to the one for which they were originally obtained. Hence, the available model by von Elsner [9] was selected since it was developed within the same study location. Other than the air temperature, this model utilizes a cloudiness factor as an important factor in the T_{sky} estimation. Thus, for all-sky conditions, T_{sky} was expressed by Eq. (1) [9]:

$$T_{sky} = \left\{1.2\, T_o - 21.4 + C\left(20.6 - 0.26 T_o\right)\right\} + 273.15 \qquad (1)$$

Sky conditions were modeled on the basis of the cloudiness factor C, which is a very important parameter in the longwave radiation exchange. Cloudiness greatly affects the magnitude of downwelling longwave radiation received at the surface of the earth. Therefore, cloudiness should be considered while modeling the downwelling longwave radiation.

A computer vision-based algorithm was developed in Halcon 11.0 (HALCON 11.0.3, 2012) which identified selected regions of interest on the weather maps and calculated the cloudiness situation at a given location, thus yielding a cloudiness factor C (**Figure 2**). Halcon is generally a comprehensive standard software for machine vision with an integrated development environment that is used worldwide.

The weather maps were obtained from the web-service Weather Online (WetterOnline). Within a given region, the weather map shows the cloud cover intensity and distribution. It also shows whether rain or snow is falling. Cloudiness influences the longwave radiation emitted by the atmosphere downward to the earth's surface.

For all-sky conditions, the downwelling longwave radiation LWR_d has the general form given by Eq. (2) [10, 11]:

$$LWR_d = \varepsilon_a\, \sigma\, T_a^4 \qquad (2)$$

The ε_a formulation has the basic structure expressed by Eq. (3) [5]:

$$\varepsilon_a = \varepsilon_{cs}\left(1 + b\; C^d\right) \qquad (3)$$

The positive relationship of the radiation with the air temperature and cloudiness indicates that empirical models can be used in the simulation under all-sky conditions.

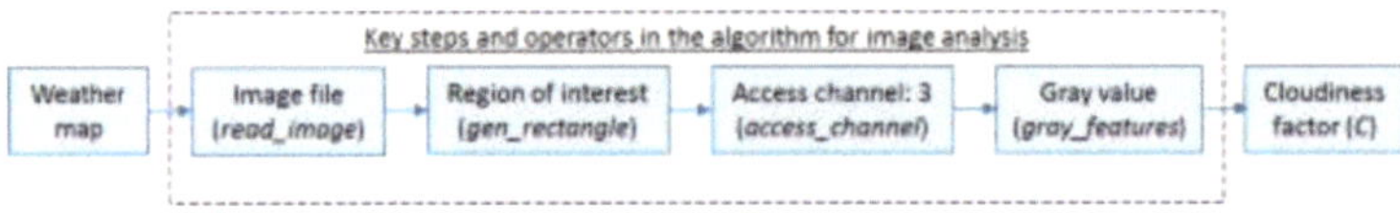

Figure 2.
Procedure overview for the image analysis with Halcon 11.0.

According to Howard and Stull [12], longwave radiation from the surrounding objects such as trees can enhance the total downwelling longwave radiation $LWR_{d,t}$ and should not be neglected. This is specifically added for comparison with the measurement from the net radiometer. $LWR_{d,t}$ is therefore expressed by Eq. (4) as:

$$LWR_{d,t} = LWR_d + \varepsilon_{gnd} F_{gnd} \sigma T_a^4 \tag{4}$$

The view factor gives the fraction of the view from a base surface obstructed by a given other surface [2]. It can be calculated numerically or analytically. A horizontal surface can see the whole sky, hence it radiates to the whole sky and its view factor with respect to the sky is equal to one. For a non-horizontal surface (e.g. roof and wall), the view factor has to be used since this is less than one. A vertical surface (tilt angle from the vertical plane = 0°) will only see half of the sky. The radiation that leaves the inclined surface is either incident on the ground or it goes to the sky (**Figure 3**).

An additional term accounting for the reflected downwelling radiation is incorporated in the computation of the upwelling longwave radiation [13]. From the equations above, the sum of the emitted longwave radiation by the surface LWR_u and the reflected downwelling longwave radiation gives the total upwelling longwave radiation $LWR_{u,t}$ [14]. Generally, the upwelling longwave radiation LWR_u can be computed once the surface temperature T_s and emissivity ε_s are known. The difference in all upwelling radiation and all downwelling radiation must result in $Q_{s,eff}$. Thus the $LWR_{u,t}$ is expressed in the form given by Eq. (5):

$$LWR_{u,t} = LWR_u + (1 - \varepsilon_s)\, LWR_d = Q_{s,eff} + LWR_d \tag{5}$$

Prediction models provide a more realistic understanding of the thermal radiation exchange between the greenhouse surfaces and the sky if all the required parameters can be accurately determined. The clear-sky atmospheric emissivity parameterizations that include both the near-surface water vapor pressure and the air temperature tend to outperform those consisting of only the air temperature.

Considering an exterior surface and the relevant parameters, the thermal radiation exchange at the surface Q_s is the sum of the components due to the exchange with the sky, the air and the ground, as expressed in Eq. (6):

$$Q_s = \varepsilon_s \sigma \left\{ \varepsilon_{sky} F_{sky} \left(T_s^4 - T_{sky}^4\right) + \varepsilon_a F_{air} \left(T_s^4 - T_a^4\right) + \varepsilon_a F_{gnd} \left(T_s^4 - T_{gnd}^4\right) \right\} \tag{6}$$

2.2 Solar irradiance on tilted surfaces

According to El-Sebaii et al. [15], estimation of total solar radiation incident on tilted surfaces can be expressed by Eq. (7) as:

$$I_{t,t} = I_{b,h}\, \psi_b + I_{d,h}\, \psi_d + I_{g,h}\, \rho_g\, \psi_r \tag{7}$$

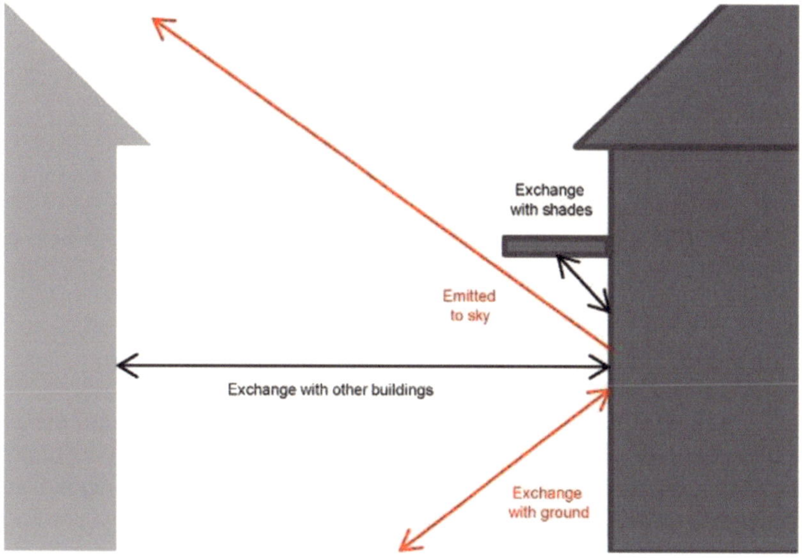

Figure 3.
Longwave radiation exchange processes at an exterior building surface.

The radiation conversion factors (Ψ_b, Ψ_d and Ψ_r) are useful in transforming the horizontal solar radiation components (**Figure 4**) to compute the total solar irradiance on the tilted surfaces.

For a surface with a given orientation, the daily value of Ψ_b is related to the time variation of incident beam radiation, the intensity of which on the ground level is a function of the atmospheric transmittance. With an angle of incidence ϑ, a zenith angle ϑz and an inclination angle β, the radiation conversion factors are given by Eqs. (8)–(10) [15, 16]:

$$\psi_b = \frac{\cos \theta}{\cos \theta_z} \tag{8}$$

$$\psi_d = \frac{1 - \cos \beta}{2} \tag{9}$$

$$\psi_r = \frac{1 + \cos \beta}{2} \tag{10}$$

2.3 Exterior surface energy balance

The energy balance at the exterior greenhouse surface is necessary in order to establish the net radiation gain (daytime solar gain) or the net radiation loss (due to heating at night). The net radiation R_n is important for surface energy analysis and is generally defined as the difference between incoming and outgoing radiation of both short and long wavelengths [17]. This net (all-wave) radiation R_n at the surface can be determined as the algebraic sum of the net shortwave radiation $R_{n,sw}$ and the net longwave radiation $R_{n,lw}$ [18], given by Eq. (11):

$$R_n = R_{n,sw} + R_{n,lw} \tag{11}$$

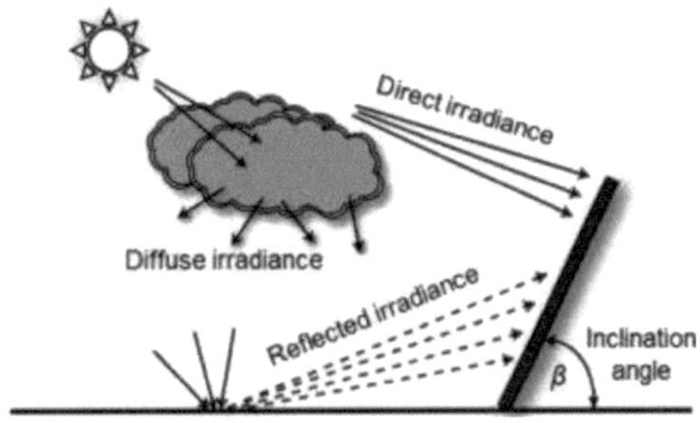

Figure 4.
Components of solar irradiance on tilted surfaces.

This net radiation balance R_n considers the total solar irradiance and the reflected component for $R_{n,sw}$, while the downwelling and the upwelling longwave radiation components are used in the calculation of $R_{n,lw}$. Hence, the R_n is further expressed as (Eq. (12)):

$$R_n = I_{t,t}\left(1-\alpha_s\right) + LWR_{d,t} - LWR_{u,t} \tag{12}$$

During the day, the sun which generally provides a large amount of radiation assures a net gain of energy, because the losses are much smaller. This net gain of energy causes a subsequent greenhouse air temperature rise. However, at night, the warm masses within the greenhouse (earthen floor, concrete paths, metal benches, plants, etc.) produce significant radiation losses to the colder outdoor environment. The net energy loss is caused by the transmission of infrared and thermal radiation through the cover, as well as the emission of radiation from the cover to the cold sky.

Under daytime and nighttime situations, the net radiation of the greenhouse is important for the evaluation of the greenhouse energy situation. It is essentially a measure of the fundamental energy available at greenhouse surfaces. A combination of night sky conditions (e.g. cloudiness, atmospheric emissivity, relative humidity) and the location of adjacent surfaces (such as other greenhouses or buildings) can directly affect the net radiation losses. For a dry greenhouse system (with no plants), energy balance requires knowledge of air exchange rate.

3. Greenhouse surface inclination and orientation

Four identical thermal boxes were developed to represent the surfaces of a glass-covered greenhouse. The four boxes were necessary in order to achieve the east, west, north, and south orientations, while changing the inclination angles characterizing the standard Venlo greenhouse surfaces. Each of the developed thermal boxes measured 1.2 m long, 0.95 m wide and 0.6 m high. The base and sidewalls of the boxes were made of Styrodur (BASF, Germany) with a thickness of 10 cm and a lightweight construction. The Styrodur also has excellent insulation properties, high compressive strength, low water absorption and resistance to aging and decay. The initial determination of the air exchange rate due to leaks with a tracer gas [19] proved that the boxes were identical. The errors due to workmanship and closing of the boxes were therefore minimized as much as possible. The exterior surfaces were inclined such that they characterize the roof slope and the walls. Based on the revised German standard for Venlo greenhouses, the roof had an inclination angle of 24° [20]. As expected, both the side and end walls of the Venlo-type greenhouse design had an angle of 90°.

This approach enabled a proper evaluation of the variations in key parameters at the external surfaces due to varied inclination and orientation. In order to avoid obstructions from buildings and trees, an appropriate rooftop was selected for positioning of miniaturized thermal boxes for assessing surface inclination and orientation effects on thermal radiation exchange (**Figure 5**). The measured parameters included net radiation, air temperature, inside and surface temperatures of the boxes, and wind speeds at different directions [7].

A window heating pad (ProfiPower, axhess GmbH & Co. KG, Hausen, Germany) was attached to the bottom section inside the thermal boxes. It was provided with 12 V DC power and in return supplied about 120 W (10 A, 12 V). The heating pad measured 40 cm by 100 cm and weighed about 0.6 kg. The maximum temperature attained by the heating pad was 55 ± 5°C and it had an integrated thermostat for temperature control. A switch-mode DC power supply unit (model 6459, Graupner GmbH & Co. KG, Kirchheim/Teck, Germany) was used. The input voltage was 230 V while the output voltage varied between 5 V and 15 V. The output current was adjustable in the range of 0 A to 20 A. Adjustment of the voltage and ampere knobs gave the needed voltage and current values, respectively. In order to reduce the voltage drop, each DC power supply unit was connected to the heating pad using a twin wire cable of 6 mm^2 cross-sectional area and approximately 46 m length. To ensure uniform heat distribution within the box, an aluminum sheet was attached firmly to the upper side of the heating pad. The aluminum sheet was 0.98 m long, 0.65 m wide, and 0.003 m thick.

During the measurement period (October 2014 to March 2015), temperature regulation was necessary to ensure that the inside temperatures in all the four thermal boxes were similar at any given time. This regulation was done with the ProfiLab Expert 4.0 program by setting the inside temperature T_i to 8 K above the ambient air temperature T_a. The program ensured that the heating pad in the boxes remained heated whenever the interior air temperature dropped below the set-point. With an output current of approximately 8 A from the DC power

Figure 5.
An arrangement of thermal boxes for surface radiation measurement.

supply, four modular monostable DIN relays (22 Series DPST-NO, FINDER GmbH, Trebur-Astheim, Germany) were connected in between the ME-UBRE relay box (Meilhaus Electronic GmbH, Alling, Germany) and the power supply units. The DIN relays used are equipped with 20 A, 250 V AC contacts rated at 5000 VA AC1 and are ideal for use in commercial applications including heating, air conditioning, and lighting. They were also suitable for this regulation since their operating temperature range is −40–40°C.

3.1 Surface inclination effects

The variation of surface-to-air temperature difference $\Delta T_{s\text{-}a}$ for both the roof and the wall in the four selected wind speed classes are compared in **Figure 6**. The box plots in each of the four directions (south, west, north, and east) display the variability of $\Delta T_{s\text{-}a}$ as the wind speed increases. For all the box orientations, $\Delta T_{s\text{-}a}$ declined with an increase in wind speed. This trend further shows that the wall $\Delta T_{s\text{-}a}$ was always higher than the roof $\Delta T_{s\text{-}a}$ and this was apparently not influenced by the directions of the thermal boxes [7]. Further tests through an analysis of variance (ANOVA) confirmed an insignificant effect of the orientation on the $\Delta T_{s\text{-}a}$ trend ($p > 0.05$).

Unlike in the big south-facing thermal box where the view factor remained unchanged, the case was different in the miniaturized thermal boxes. By changing the inclination angle β, the view factor is similarly altered [2]. In this respect, when β changes from 24° (roof) to 90° (wall), the view factor of the exterior surface to the sky F_{sky} is reduced while that to the surrounding ground objects F_{gnd} is increased. At an inclination angle of 24°, the roof is the most exposed component of the greenhouse structure. This in turn led to lower surface-to-air temperature differences $\Delta T_{s\text{-}a}$ of the roofs compared to those of the walls. This implies that the sky-oriented exterior roof surfaces are cooled more than the vertical walls.

However, under an overcast condition, the variation in inclination angles did not show any significant changes ($p > 0.05$) in the net longwave radiation loss. This supports the fact that the radiative heat flux is not well connected to the surface inclination as it merely depends on the temperature difference. As expected, the exposed roof loses more heat to the sky than the walls under clear-sky conditions. The surface of interest represents that of the Venlo greenhouse design where the roof fraction is low. This outcome agrees well with the observation that the night-time heat loss by longwave radiation affects any building surface whose roof fraction is high. This is equivalent to saying that the sensible heat flux is higher when the roof area is more than the wall area. The reduced roof surface area (an area of

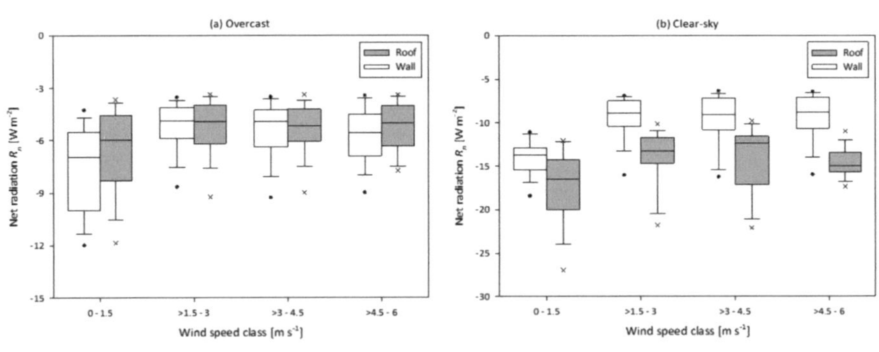

Figure 6.
Variation of mean net radiation at roof and wall surfaces with wind speed under all-sky conditions: (a) overcast (n = 35 nights), and (b) clear-sky (n = 6 nights).

Surface orientation	Selected wind speed classes [m s^{-1}]			
	0–1.5	> 1.5–3	> 3–4.5	> 4.5–6
South	1.60 [a]	1.49 [b]	0.75 [c]	0.58 [e]
West	1.82 [a]	1.53 [b]	1.14 [d]	1.01 [f]
North	1.63 [a]	1.30 [b]	1.22 [d]	1.13 [f]
East	1.83 [a]	1.41 [b]	1.15 [d]	1.09 [f]
Mean ± Stdev	1.72 ± 0.12	1.41 ± 0.15	1.07 ± 0.22	0.95 ± 0.26
(Within column, same letter indicates insignificant differences at 5% level)				

Table 1.
Deviation between wall and roof surface-to-air temperature difference values for different orientations and wind speeds.

major heat loss) in Venlo greenhouses is therefore beneficial in the overall reduction of the heating costs.

3.2 Surface orientation effects

For the chosen wind speed classes and surface orientations, the deviation between the surface-to-air temperature difference $\Delta T_{s\text{-}a}$ of the wall and that of the roof was represented by $\Delta T_{W\text{-}R}$. The values of $\Delta T_{W\text{-}R}$ (in K) are given in **Table 1**. The mean $\Delta T_{W\text{-}R}$ was highest at low wind speed (< 1.5 m s^{-1}) and lowest at high wind speed (> 4.5 m s^{-1}). Interestingly, the standard deviation (Stdev) increased with an increase in wind speed, with a range of 0.12 K to 0.26 K. Despite the random orientation of the thermal boxes, the variation in $\Delta T_{W\text{-}R}$ within the same wind speed class did not show a significant difference ($p > 0.05$). For wind speeds ≥3 m s^{-1}, the south-facing surface registered the lowest values in terms of deviation in $\Delta T_{s\text{-}a}$ unlike the other three surface orientations [7].

Southerly and westerly wind directions were generally dominant during the measurement period. It is also worth noting that the dominant wind speed class was that between 1.5 m s^{-1} and 3 m s^{-1}. It was also apparent that wind was very variable both in direction and speed. An increase in wind speed reduces the surface resistance; this generally leads to an increased heat loss which is largely brought about by convection.

Based on the trends of the surface-to-air temperature difference $\Delta T_{s\text{-}a}$, the deviation between the $\Delta T_{s\text{-}a}$ values of the roof and the wall (i.e. $\Delta T_{W\text{-}R}$) was not significantly affected by the box orientation. Interestingly, this effect of orientation on nighttime R_n from this study was insignificant. This indicates that the R_n was little affected by varying the orientation of the thermal boxes. Generally, these orientations are applicable, especially during the day, in maximizing winter sunlight and heat gain depending on whether the greenhouse is a single-span or a gutter-connected type [3].

4. Conclusions

At the greenhouse surfaces, the weighted contributions of thermal emissions from the sky, the surrounding air, and the ground objects are explained by the view factors. During a clear night, the greenhouse surface loses more heat as it radiates to the very cold clear sky. On a regional scale, clouds play a critical role in the radiation balance at the surface. Under both day and night situations, the study delivers reliable results in terms of the calculation of parameters necessary for the radiation models. The parameters which have an influence on the daytime and nighttime net radiation are surface emissivity, atmospheric emissivity, surface and atmospheric temperatures, and albedo.

With reference to surface inclination and orientation effects, the findings of the study are useful in understanding the impacts of the variously inclined and oriented greenhouse surfaces on heating energy and thus on heat losses. Furthermore, it is important to consider the impact of wind speed specifically for the windward greenhouse surfaces in energy simulations. In this case, the data is distinguished from those of leeward surfaces and the sensitivity to the variation in wind direction can be checked. This becomes more important in uncertainty quantification as a result of variations in the surface orientation.

Acknowledgements

The author is grateful for the joint scholarship support (between the National Commission for Science, Technology and Innovation (NACOSTI), Kenya and the German Academic Exchange Service (DAAD), Germany) and the material support from Biosystems Engineering Section, Gottfried Wilhelm Leibniz Universität Hannover, Germany.

Conflict of interest

The author declares that there is no conflict of interest regarding the publication of this chapter.

Nomenclature

b, d	constants determined experimentally [–]
C	cloudiness factor [–]
F_{sky}	view factor to the sky [–]
F_{air}	view factor to the air [–]
F_{gnd}	view factor to the ground [–]
$I_{b,h}$	beam radiation [W m^{-2}]
$I_{d,h}$	diffuse radiation [W m^{-2}]
$I_{g,h}$	global radiation on a horizontal surface [W m^{-2}]
$I_{t,t}$	total solar radiation on the tilted surface [W m^{-2}]
LWR_d	downwelling longwave radiation [W m^{-2}]
$LWR_{d,t}$	total downwelling longwave radiation [W m^{-2}]
LWR_u	upwelling longwave radiation [W m^{-2}]
$LWR_{u,t}$	total upwelling longwave radiation [W m^{-2}]
Q_s	thermal radiation exchange [W m^{-2}]
$Q_{s,eff}$	effective thermal radiation exchange [W m^{-2}]
R_n	net (all-wave) radiation [W m^{-2}]
$R_{n,sw}$	net shortwave radiation [W m^{-2}]
$R_{n,lw}$	net longwave radiation [W m^{-2}]
T_a	air temperature [K]
T_{gnd}	ground temperature [K]
T_s	surface temperature [K]
T_{sky}	sky temperature [K]

α_s	albedo of the earth surface [−]
$\Delta T_{s\text{-}a}$	surface-to-air temperature difference [K]
$\Delta T_{W\text{-}R}$	deviation between wall and roof $\Delta T_{s\text{-}a}$ values [K]
ε_{gnd}	ground emissivity [−]
ε_a	effective atmospheric emissivity [−]
ε_{cs}	clear-sky atmospheric emissivity [−]
ε_s	surface emissivity [−]
ε_{sky}	sky emissivity [−]
Ψ_b	beam radiation conversion factor [−]
Ψ_d	diffuse radiation conversion factor [−]
Ψ_r	ground reflected radiation conversion factor [−]
ρ	angle of inclination from horizontal [°]
ρ_g	ground reflectivity [−]
σ	Stefan-Boltzmann constant = 5.67×10^{-8} [W m^{-2} K^{-4}]
θ	angle of incidence [°]
θ_z	zenith angle [°]

Author details

Erick K. Ronoh
Department of Agricultural and Biosystems Engineering, Jomo Kenyatta University of Agriculture and Technology, Nairobi, Kenya

*Address all correspondence to: ronoh@jkuat.ac.ke

References

[1] Vollebregt HJM, van de Braak NJ. Analysis of radiative and convective heat exchange at greenhouse walls. Journal of Agricultural Engineering Research. 1995;**60**(2):99-106. DOI: 10.1006/jaer.1995.1004

[2] Evins R, Dorer V, Carmeliet J. Simulating external longwave radiation exchange for buildings. Energy and Buildings. 2014;**75**:472-482. DOI: 10.1016/j.enbuild.2014.02.030

[3] Sanford S. Reducing greenhouse energy consumption: An overview. Energy Efficiency in Greenhouses, Series A3907-01. University of Wisconsin-Extension, Cooperative Extension. Madison, WI, USA; 2011. 15 p.

[4] Gupta MJ, Chandra P. Effect of greenhouse design parameters on conservation of energy for greenhouse environmental control. Energy. 2002;**27**(8):777-794. DOI: 10.1016/S0360-5442(02)00030-0

[5] Duarte HF, Dias NL, Maggiotto SR. Assessing daytime downward longwave radiation estimates for clear and cloudy skies in southern Brazil. Agricultural and Forest Meteorology. 2006;**139**(3-4):171-181. DOI: 10.1016/j.agrformet.2006.06.008

[6] Araújo AL, da Silva BB, Braga CC. Simplified modeling of downwelling long-wave radiation over Brazilian semi-arid under irrigation conditions. Brazilian Journal of Geophysics. 2012;**30**(2):137-145. DOI: 10.22564/rbgf.v30i2.87

[7] Ronoh EK, Rath T. Effects of greenhouse surface inclination and orientation on exterior longwave radiation exchange. Journal of Sustainable Research in Engineering. 2017;**3**(4):105-112

[8] Ronoh EK, Rath T. Modelling of longwave radiation exchange at greenhouse surfaces under all-sky conditions. Agricultural Engineering International: CIGR Journal. 2015;**17**(4):23-35

[9] Von Elsner B. Das Kleinklima und der Wärmeverbrauch von geschlossenen Gewächshäusern: Ein Simulationsmodell zur gartenbautechnischen Bewertung unter Berücksichtigung des Einflusses von Standortklima, Pflanzenbestand und Gewächshauskonstruktion. PhD Thesis, Gartenbautechnische Informationen. In: Heft. Vol. 12. Germany: Institut für Technik in Gartenbau und Landwirtschaft, Universität Hannover; 1982

[10] Choi M, Jacobs JM, Kustas WP. Assessment of clear and cloudy sky parameterizations for daily downwelling longwave radiation over different land surfaces in Florida. USA. Geophysical Research Letters. 2008;**35**(L20402):1-6. DOI: 10.1029/2008GL035731

[11] Dos Santos CAC, Da Silva BB, Rao TVR, Satyamurty P, Manzi AO. Downward longwave radiation estimates for clear-sky conditions over northeast Brazil. Revista Brasileira de Meteorologia. 2011;**26**(3):443-450. DOI: 10.1590/S0102-77862011000300010

[12] Howard R, Stull R. Modeling the downwelling longwave radiation over a groomed ski run under clear skies. Journal of Applied Meteorology and Climatology. 2013;**52**(7):1540-1553. DOI: 10.1175/JAMC-D-12-0245.1

[13] Tang BH, Li ZL. Estimation of instantaneous net surface longwave radiation from MODIS cloud-free data. Remote Sensing of Environment. 2008;**112**(9):3482-3492. DOI: 10.1016/j.rse.2008.04.004

[14] Liang SL. Quantitative remote sensing of land surfaces. New Jersey, USA: John Wiley & Sons; 2004 560 p

[15] El-Sebaii AA, Al-Hazmi FS, Al-Ghamdi AA, Yaghmour SJ. Global, direct and diffuse solar radiation on horizontal and tilted surfaces in Jeddah, Saudi Arabia. Applied Energy. 2010;**87**(2):568-576. DOI: 10.1016/j.apenergy.2009.06.032

[16] Ronoh EK. Prediction of total solar irradiance on tilted greenhouse surfaces. Agricultural Engineering International: CIGR Journal. 2017;**19**(1):114-121

[17] Choi M. Parameterizing daytime downward longwave radiation in two Korean regional flux monitoring network sites. Journal of Hydrology. 2013;**476**(1):257-264. DOI: 10.1016/j.jhydrol.2012.10.041

[18] Ayoola MA, Sunmonu LA, Bashiru MI, Jegede OO. Measurements of net all-wave radiation at a tropical location, Ile-Ife. Nigeria. Atmósfera. 2014;**27**(3):305-315

[19] Tantau H-J. Heat requirement of greenhouse including latent heat flux. Landtechnik. 2013;**68**(1):43-49

[20] Von Elsner B, Briassoulis D, Waaijenberg D, Mistriotis A, von Zabeltitz C, Gratraud J, et al. Review of structural and functional characteristics of greenhouses in European Union countries, Part II: Typical designs. Journal of Agricultural Engineering Research. 2000;**75**(2):111-126. DOI: 10.1006/jaer.1999.0512

Chapter 5

Greenhouse Requirements for Soilless Crop Production: Challenges and Prospects for Plant Factories

Aliyu Idris Muhammad, Abubakar Shitu, Umar Abdulbaki Danhassan, Muhammad Hilal Kabir, Musa Abubakar Tadda and Attanda Muhammed Lawal

Abstract

This chapter discussed the greenhouse requirement for soilless crop production. It further introduced soilless crop production and elucidated the equipment required for an efficient production system covering greenhouse environmental control and management of temperature, humidity, lighting, and nutrients using innovative strategies. Also, the energy required for the control of the greenhouse environmental conditions during the crop production cycle was explained. Identification and management of pests and diseases using wireless network sensors and the Internet of Things for efficient and safe food production were also highlighted. Finally, the challenges facing greenhouse crop production itemized, and the prospects of greenhouse technology for sustainable healthy food production were proposed.

Keywords: greenhouse crop production, hydroponics, greenhouse energy requirement, pest management, wireless network sensors

1. Introduction

Greenhouse crop production is an agricultural management technique employed nowadays for increasing food production under a controlled environment. It is an emerging, efficient, and feasible alternative guaranteeing food supply throughout the year without any hindrance from the external environmental factors. In recent years, technological advancements such as wireless sensor networks and agricultural robots have been able to handle the challenges facing greenhouse farming by overcoming its limitations, mitigating adverse impacts from the climate and environmental changes, and ensuring system sustainability [1, 2]. In greenhouse crop production, two methods of farming crops are employed; namely, crop production that involves soil as the growing medium and soilless crop production popularly referred to as hydroponics, which utilizes nutrients in a liquid medium.

Soilless crop production refers to any technique of growing crops in the absence of soil as a rooting medium to boost the yield, quality, and safety of food products that meet the demands of consumers. The main advantage of this method is reducing the problems associated with soil, including soil salinity, poor soil structure and quality, soil-borne pests and diseases, and non-arable soil [3]. For soilless crop production in the greenhouse, one has the leverage to control many limiting factors that crops encounter, including temperature, light, and a large degree of pest and disease problems, and soil-related problems mentioned above. Moreover, soilless crop production in the greenhouse can be practice all year-round, including during the winter months, where good-quality crops are guaranteed. This chapter will discuss soilless crop production in the greenhouse under the following sub-sections.

2. Greenhouse production system

In recent years, the agricultural sector has witnessed a rapid increase in the use of the greenhouse production system (GHPS) as an alternative to the growing demand for food around the globe. The controlled environment in the greenhouse system guarantees food safety and high crop yield in limited space, especially in populated areas. With this closed-field cultivation, simple rows of vertical or open-fields crops are cultivated in nutrient media under a highly controlled environment. The advent of smart farming (precision technology, wireless sensors, and data processing) is changing the crop cultivations from conventional greenhouses to advanced high-tech plant factories for the optimization of human labor and boosting crop productivity. High-tech greenhouse production systems such as the one shown in **Figure 1** are also refered to as controlled environment plant production systems, controlled environment agriculture, or phytomation systems [4].

Traditionally, GHPS uses natural or artificial light to optimize growth conditions of horticultural crops, fruits, and vegetables, or plant research programs, thereby reducing the food threat projected by the United Nations. However, there is a need to promote scientific solutions that can lead to more efficient production of crops in the greenhouse via optimization of various environmental conditions and subsystems and the understanding of the external factors that should be integrated into the system. In this way, the technical aspects of the GHPS, including automation, culture, and environment, need to be integrated so that the system and its goals could lead to a conclusion regarding the system's performance indicators. These cultural and environmental factors comprise crop cultivation procedures based on plant physiognomies, growth responses, and microclimate requirements. It also constitutes physiological, planting, and post-harvest processes, harvesting, and packaging. The strategies for using affordable and energy-saving facilities as

Figure 1.
View of a high-tech greenhouse facility in the Netherlands (source: Meteor systems BV).

covering materials and lighting and microclimate control systems are necessary for a viable GHPS. This will reduce the cost of automated energy management and environmental impact and maximize the use of natural resources.

2.1 Greenhouse cover materials

Although GHPS is widely accepted in Europe and China, the initial cost seems to be the main factor limiting its acceptability to low-income farmers, especially in developing countries that needed the technology the most. The materials used to support foundation, shape, and framings for establishing a geographical direction and optimal light entrance add to the high initial cost of GHPS. Greenhouse structures and covering enclose the cultivation area and space. Transparent materials such as polyethylene (PE) films, ultraviolet stabilized PE-films, and numerous transmitting cover materials were reported elsewhere [5]. A study found that the combination of PE cover and silicon double glazing photovoltaic panels reduced solar radiation by 35–40% as against PE cover alone. The researchers further observed that the silicon double glazing photovoltaic panels shading reduced the air temperature of the protected cultivation of tomatoes and peppers crops. Screen-houses are often used in greenhouse operating on natural ventilation in the tropical lowlands regions to control insects, intense solar radiation, risk of heavy rainfall, and strong wind [6]. A comprehensive review of the challenges with the use of photovoltaic panels in the greenhouse crop production system is available in [7]. Insect-proof nets are employed to cover the greenhouse and maintain the inside and outside temperatures at a required level.

On the other hand, photo-selective films increase the temperature during the summer. Studies showed that greenhouses covered with net-screen are more prevalent in tropical regions because of the climate conditions that have optimality degrees near the plants' desired levels. Greenhouses covered with insect-proof net-screens and operate on natural ventilation have their internal and external air temperatures maintained at the same level [4]. Also, shading nets can protect plants from excessive sunlight, heavy rains, and wind and facilitate the natural ventilation process. What is now needed is to synthesize these materials from renewable source materials.

2.2 Light control in greenhouse systems

Light control and interception of radiance in GHPS are by using shading screens, planting density, and artificial lights [6]. Since light conditions and air temperature are the most critical environmental factors for plant growth. Analyzing optimal air temperature without discussing plant evapotranspiration and the light condition does not generate any useful data to maximize productivity and high-quality yield. The correlation between light and air temperature is high such that one cannot be optimized without consideration of the other. For instance, the quality of tomato (yield, lycopene content, and productivity) is not only influenced by the microclimate parameters and cultural practice but also by the photosynthetic photon flux density. The photosynthetic photon flux density is a condition of the optimal combination of light, relative humidity, and air temperature, resulting in maximum yield.

3. Requirements of a modern greenhouse for crop production

Greenhouse production systems demand an efficient strategy to control microclimate conditions, including humidity, temperature, and gas level, to maintain an optimum ambient setup for crop cultivation [8]. One of the most important technologies of the 21st century is the wireless sensor network (WSN), which is very suitable for distributed data collection and monitoring in complex environments such as greenhouses. Many measurement points are needed to trace down the local climate settings in various locations of the greenhouse to ensure proper operation and automation of the production cycle. Cabling this measurement operation is expensive, vulnerable, and challenging to relocate once installed [9].

In emerging greenhouses, technological advancement such as WSN has brought solutions in precision agriculture (PA). Modern agricultural management practices require WSN-enabled equipment to efficiently manage the various microclimate parameters to achieve high-quality agricultural produce. These WSN gadgets utilize Low-Power Wide-Area Networks (LPWANs) as a wireless technology for long-distance data transmission with minimal power consumption. Because greenhouses are liable to several changes and interference, they require a better WSN design scheme to manage and process data. Long Range Wide Area Network (LoRaWAN) is a low data rate and among the most successful LPWAN technologies used nowadays due to its low deployment and management costs [10].

Greenhouse operations depended on the technologies employed in the covering materials, structure orientation, shape, dimensions, and microclimate control [11]. Modern greenhouses are faced with challenges due to intricate structural design. The other challenges encountered in the greenhouse include design adaptation with respect to different crops, the impact of the metal structures, and the technology employed for PA. These challenges have a significant influence on the growth of crops, which necessitated an adaptive precision monitoring solution. WSNs provide an effective solution in managing greenhouses, with an efficient strategy in different fundamental aspects of sensors types, connectivity, network optimization, and power source [12]. Meanwhile, LPWANs are low power consumption and long-range communication gadget suited for wireless communication in greenhouse PA [13]. However, there are some crucial challenges in WSN deployment that have restricted its real benefits in PA, including maintaining coverage, optimal deployment scheme, long-range connectivity for communication, and energy-efficient network for extended battery life [14].

Nowadays, as the internet has become inevitable in daily life, all devices need a network to function and communicate with other devices. This is brought about by the Internet of Things (IoT), which is another promising technology of the 21st century that finds excellent applications in farming and makes the life of a farmer easier. IoT is used in the connection, control, and management of intelligent devices connected to the internet [15]. The technology enables people to access different data over the internet from any remote location. It is considered the third wave of information technology after the internet and mobile communication network, with more intelligence and comprehensive interoperability [16]. Thus, numerous sensors and controllers are used in collecting environmental data in a greenhouse and send it to the control station over the internet.

Different seasonal crops are grown only under certain conditions. Onions, garlic, and shallots are winter crops that require cold conditions for their growth, whereas cucumbers and melons are summer crops that require moderate or hot climatic conditions. IoT and various sensors help bring solutions to many of the existing practical problems over the years [16]. Numerous sensors and controllers have been utilized in greenhouse environmental data collection and send such data to the control station over the Internet. Therefore, three crucial factors of WSN are considered during operation, namely; power consumption, the accuracy of measurement, and network connectivity. Besides, there are other considerable aspects of WSN, including stability, cost, and data security that are of importance [17, 18].

4. Management and control of pests in the greenhouse

4.1 Some greenhouse pests

Various insect pests and other arthropods attack greenhouse crops during the soil or soilless cultivation. These insect pests are categorized into chewing insects and sucking arthropods depending on the type of attack and damage they cause to crops. The chewing insect pests of greenhouse crops mostly feed on leaves and plant roots. The larvae of these insects include Lepidoptera, Coleoptera, Diptera, and Hymenoptera. Others are larvae of fungus gnats that tunnel into the stems of many crops or feed on the roots of plants. All these cause severe damages to seedlings and cuttings, whereas large crops are minimally impacted. Control of such infestations is problematic in the greenhouse due to the over-lapping and tunneling behavior of these pests [19].

On the other hand, sucking arthropod pests in greenhouse crops feed on host plants by piercing and sucking liquid biomaterial from the plants, thereby damaging their tissues. These type of pests used their needle-like mouthparts to penetrate the plants' tissues. They include many pest species such as Western flower thrips, mealybugs, *Bemisia tabaci* (Genn.) scales, aphids, mites, and whiteflies [19, 20]. Damages caused by these pests include feeding on new leaf clusters, leaf undersides, or in developing flowers of mature plants. The pests mostly hide and protect themselves from direct pesticide spray applications and pierce and suck liquid contents underneath the epidermal cells, thereby developing pale or yellowish-white leaves, a condition of insufficient chlorophyll and with visible fruits deformities in some plants. Some pests attack greenhouse vegetables and ornamental crops by feeding on phloem-sap, thereby excreting abundant excess water and sugars that encourage molds growth and transmission of plant viruses. Some sucking pests (aphids) damage greenhouse crops by causing stunted and abnormal growth in host crops by injecting toxic substances [19].

4.2 Pest control in greenhouse

4.2.1 Pests control using conventional methods

The control of pests in the greenhouse is a critical management practice to ensure food safety, quality, and bumper harvest. In greenhouse crop production, control measures are essential in soilless crop cultivation since the greenhouse provides an excellent environment for different pests to thrive [21]. The prevention of pest attacks on crops is much easier than trying to control the infestation after the attack. The control and management of pests practice include biological, chemical, and physical methods. In the physical method, light has been employed to attract insect pests to control devices or adjust light wavelengths to control insects and diseases [21, 22]. Nowadays, biological control is preferred over chemical control due to health concerns. El Arnaouty et al. [20] reported that the application of the biological control method, such as the use of different parasitoids and predators, has yielded an appreciable safe production of sweet pepper in the greenhouse as compared with the recommended chemical control method during the same period. The biological control method has increased the yield of sweet pepper by 35.06 and 17.88% as against the untreated and chemical control method, respectively [20].

Geographical location, crop type, and method of crop cultivation determine the type of pest that can be found in a particular greenhouse. Therefore, specific control measures need to come in handy to farmers and familiarize themselves with the pest

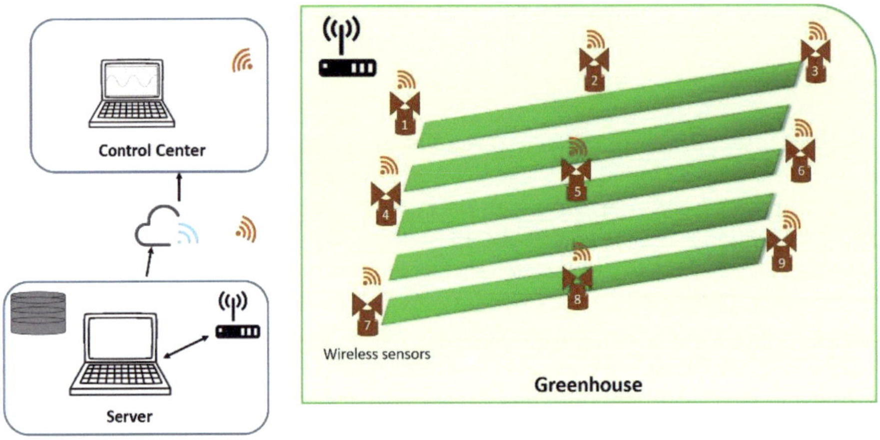

Figure 2.
Wireless sensor network for greenhouse management of crop pests.

in their locality and control measures to avoid damaging the crops. Fungicides and pesticides recommended for such an area should be available as at when needed in a timely manner to avoid economic catastrophe if pests are not urgently handled. Farmers should know all the levels of pest incidences by conducting daily routine monitoring to allow them to distinguish at which level a pest can be tolerated that might not require treatment [23].

4.2.2 Pests control using smart approaches

With the recent advancement in smart agriculture, noninvasive approach such as the use of sensors, Unmanned Aerial Vehicles (UAV), and artificial intelligent noses (electronic noses) have been deployed in open farms and greenhouses for real-time monitoring, identification, and control of crop pests and diseases [24, 25]. Different types of sensors employed for pests detection include gas detection, sound detection, and spectral remote sensors. A typical wireless sensor network deployed for greenhouse applications is showcased in **Figure 2**.

4.2.2.1 Gas detection sensors

Gas detection sensors utilize volatile compounds released by crops as a result of external stress from pests infestation, human disturbance, or environmental factors. During pests infestation, crops produce volatile chemical compounds to the surrounding that is recorded as gas or image. Numerous samples of the volatile compounds released by plants need to be gathered after different stresses to enable the identification of pest infestation in the area. Thermal imaging is also utilized in the characterization of volatile chemical compounds released because they have a specific spectral signature [25].

4.2.2.2 Sound detection sensors

This is a farm management (FM) practice in which wireless sensors coupled with antennas are mounted in greenhouses at strategic points to pick sound waves produced by pests as they fly, chew, or mate. The farmer then records the noise levels for the given period and analyzes the data on a computer. Pest-infested areas typically have louder sound than none or less infested areas. The farmer might opt to manage the pests instantly or wait until they reach the economic control threshold. This is an effective form of detection approach utilized by both large and small scale farmers with little cost of operation and higher than average performance. The only shortcoming is that environmental conditions might influence the data gathering during storms or heavy winds [26].

4.2.2.3 Spectral remote sensors

The technique of remote sensing is deployed in the processing, characterizing, interpreting, and displaying data as images using spectral remote sensors available as low-image sensors and high-image sensors [24]. Low-image sensors consist of cameras mounted at strategic locations to capture images and send them to a control station. They only capture visible images of pests since they are low-resolution cameras. The captured images provide information on the pests population on a particular crop and estimate the overall infestation in the greenhouse. These low-image sensors are frequently used by farmers because of their low capital investment and maintenance costs. The high-image sensors, on the other hand, detect the spectral signature of each crop and record it in a spectrum beyond the human spectrum. It includes x-rays, gamma rays, infrared, and ultraviolet rays. The image data produced can either be multispectral or hyperspectral. These high-resolution images can distinguish the physical and chemical components of the crop from thousands of kilometers away [24, 26].

Before the detection of crop pests, the imaging spectrometers are pre-loaded with spectral signatures of each crop on the greenhouse. Therefore, once pests attack the crops, their spectral signatures change because the pests absorb the crops' light and force them to reflect a divergent spectral signature than the pre-loaded ones. The pests' population and their exact location on the crop and lifecycle stage are known by analysis of the images. Compared to the low-image sensors, they are better in terms of accuracy and detection of multiple types of pests and diseases. Meanwhile, imaging spectrometers are expensive and require a significant outlay for maintenance.

4.2.2.4 Fluorescence image sensing

This approach requires analyzing the chlorophyll content in a crop. The images of crop leaves are captured and compared with images of noninfested leaves. The crop infestation is detected by variation in the chlorophyll pattern. This technique can only be utilized to identify pests on crops with chlorophyll [26].

5. Energy requirement for crop production in the greenhouse

5.1 Energy requirements

With the intensification of crop production, crop yield and quality are enhanced via control of the internal environment of the greenhouse using electricity. In this case, the reduction of fuel and electricity consumption to attain an optimum growth environment constitute the main concerns of greenhouse cultivation [27]. Heating and cooling systems for environmental control consume high electrical energy in regions with scorching weather conditions, such as Saudi Arabia, which recorded around 153 Wh m^{-2} d^{-1} for cooling systems, pad, and the air circulation

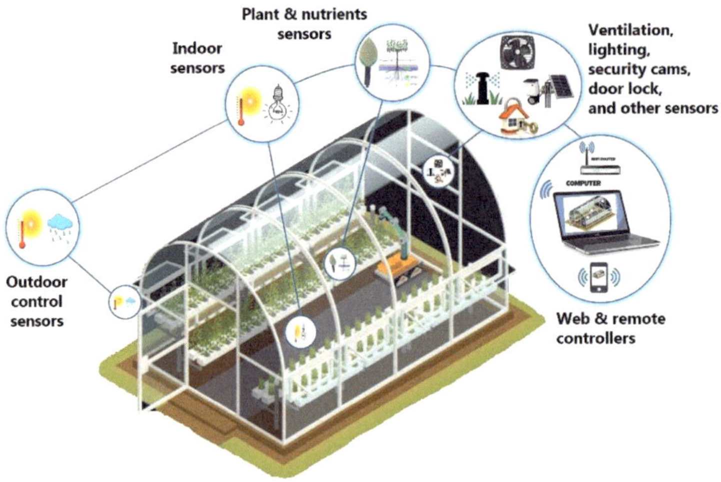

Figure 3.
A remotely-operated greenhouse for monitoring of various external and internal greenhouse environmental conditions.

fans [28]. In temperate regions, heating systems consume a large amount of electrical energy, such as in Canada, where supplementary lighting commands high electrical power, as reported by Bambara and Athienitis [29].

Moreover, the electrical energy demands of greenhouses rely on the types and operational periods of the loads, energy requirement, and efficacy of the appliances, and overall, the climatic conditions of greenhouse location. Plant species influenced electrical energy demand, and this was attributed to the respective optimal growth temperature required by different crops. For example, in Turkey, the heating demand of tomatoes, lettuce, and cucumber production in a greenhouse were 105, 29, and 217 Wh m^{-2} d^{-1}, respectively, for the maintenance of corresponding greenhouse temperatures of 20, 14, and 26°C during winter season [30]. Ntinas et al. [31] reported a substantial amount of electrical energy of 352 Wh m^{-2} d^{-1} consumption in mid-winter greenhouse during the cultivation of tomatoes in Central Macedonia. This energy consumption was recorded during heating, irrigation, and air-circulation operations [31]. Other studies stated electrical energy consumption for greenhouse cultivation of tomatoes, eggplant, cucumber, basil, and pepper as 10.2–17.1, 14.4, 9.7–29.1, 178.7, and 9.0–12.5 GJ ha^{-1}, respectively [28]. In some situations, if the costs of investments for environmental control equipment cannot be recovered from the profit gained due to the costs of fuel and electricity, it is suggested not to use the environmental control equipment in the greenhouse [28, 32]. **Figure 3** is a smart energy-saving greenhouse environmental control system for soilless crop management.

5.2 Control of greenhouse environmental conditions

Factors such as temperature, humidity, carbon dioxide, light, and other parameters influence the cultivation of crops in the greenhouse, and therefore these parameters need constant monitoring.

5.2.1 Temperature

Due to the transparent nature of roofs and walls of the greenhouse, sunlight infiltrates without hindrance. These covering materials prevent thermal leakages from the greenhouse, resulting in higher internal temperature compared to that of the outside environment. Meanwhile, in temperate climates, fuel-assisted heating and grid electricity inputs enabled the extension of the cultivation period in colder seasons. This allowed the location of greenhouses even in colder areas, leading to improve crop quality in winter. The amount of energy consumption in greenhouse crops cultivation increases with increased latitude due to heating and additional lighting [28, 33]. Greenhouse internal temperatures rise during summer in temperate regions. These high temperatures have a great impact on crop growth and development, thus, reducing economic yields [28].

5.2.2 Humidity

The humidity in the greenhouse environment has a strong influence on crop transpiration and disease infections. Plant stomata close to prevent extra transpiration in the surrounding dry air, which suppresses CO_2 exchange between the air and leaves, thus reducing the net photosynthetic rate. Therefore, managing humidity is necessary to provide an optimum environment for crop growth. Electricity is used to automatically control the environment concerning real-time variations of interior microclimate and plant conditions, which also add to the electrical energy consumption.

5.2.3 Carbon dioxide

CO_2-concentration in greenhouse fluctuates based on the respiration and photosynthesis of crops. In some instances, the concentration reaches peak levels in the morning since CO_2 generated during the nighttime respiration remained in the greenhouse. In the daytime, with insufficient ventilation and plant photosynthesis proceeds, the CO_2 level in the greenhouse declined compared to the outside environment. Ventilation devices play a critical role in maintaining the CO_2 levels in the greenhouse. Also, CO_2 supply systems are usually used for providing adequate CO_2 for sustaining crop cultivation [28].

5.2.4 Light

Sunlight is among the most important energy sources for crop photosynthesis. This energy is absorbed, transformed, and stored by crop chlorophyll molecules in the form of photon energy for their growth. It has been reported that producing 1 kg of fresh tomatoes would require an average of 6 MJ of sunlight with 90 mol of photosynthetically active radiation (PAR, 400–700 nm wavelength range) photons, which is equivalent to 900 MJ of sunlight for 1 kg dry weight of tomatoes [28]. For the cultivation of high light-demanding vegetable crops like tomatoes, the greenhouse cover design should be in such a way to deliver efficient sunlight to the crops. In Mediterranean regions, light-diffusing films are utilized to prevent leaf burning due to extreme direct sunlight on greenhouse crops. The sunlight radiation is absorbed into the crops, thus, improve the photosynthesis performance. On the other hand, complementary lighting is applied in the high-latitude regions due to the less intensity of sunlight. Also, lighting during nighttime is used to regulate flowering to enhance food product supply to markets [28, 34].

5.3 Adaptive analysis framework

Recently, computer-based microclimate control systems and simulation software for knowledge-based decision making have been deployed. Adaptive Analysis

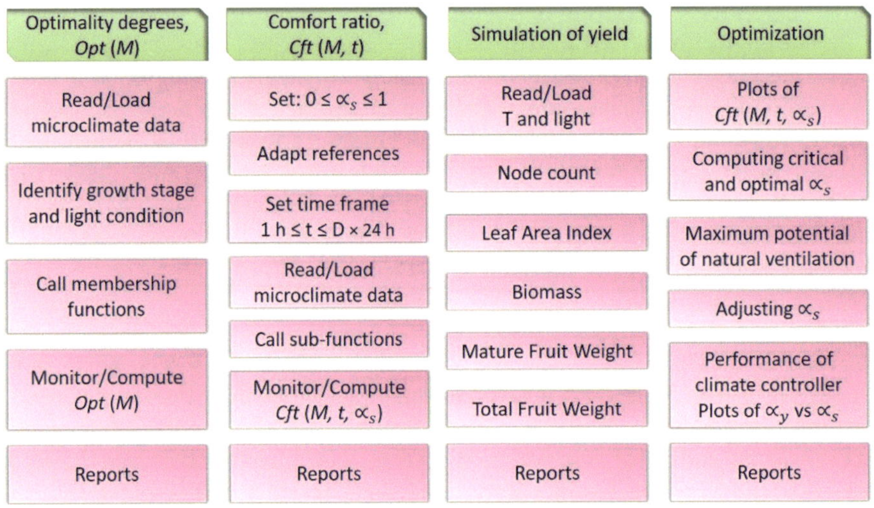

Figure 4.
An interphase of adaptive management framework toolbox redraw from [35].

framework utilizes a custom-designed data acquisition and control system [35] that has been built based on ESP32 microcontroller board for monitoring and manipulating of the microclimate parameters. Three computer models were employed by the framework for evaluation and adjusting of optimality-degrees $Opt(\mathcal{M})$, comfort ratio $Cft(\mathcal{M}, t, \alpha)$, and prediction of the expected yield as depicted in **Figure 4**. The framework was implemented in MATLAB® (The MathWorks Inc., Natick, MA, USA) environment through Simulink blocks and coding of various main functions and sub-functions that were stored as "m-files". Different toolboxes were developed for the immense data analyzing tasks. The framework structure was designed in a way that end users can create (or update) entries in database, select report type (one-day or multi-days report), and proceed with a specific analysis procedure. The database is a dynamic flat file type that can be created by entering collected data, either manually from previously stored sources such as excel sheets, or directly from the hardware interface. The computer models presented in this chapter are focused on tomato (*Lycopersicon esculentum*); however, with slight modification the framework can be reprogrammed to work with other greenhouse crops provided that their yield prediction and growth response models are available. Results of microclimate evaluation and set point manipulation generated by different Simulink blocks of the AAF can contribute to dynamic greenhouse climate control strategies [36], such as the one in [37].

6. Challenges and prospects of crop production in greenhouse

6.1 Challenges facing crop production in greenhouse

The use of greenhouses for the cultivation of crops has been described as an innovative way of mechanically and scientifically controlling the natural outdoor climatic conditions, such as torrential rainfall, high and low temperature, relative humidity, pests, and diseases [38–40]. Ventilation systems are an integral part of most greenhouses in the Mediterranean regions; thus, greenhouses in those areas are not heated at all [39]. One of the most pressing issues facing the use of greenhouse in those regions is insufficient ventilation during the summer and no heating during the winter, which causes an imbalance on the side of the optimum temperature and relative humidity conditions required for the wellbeing of the crops [39, 41]. Hence, the conception and implementation of air conditioning systems for the direct control of the climate, especially the most important variables (temperature and humidity), are inevitable [38].

Moreover, mimicking an ideal and real environment by ensuring indoor ventilation and other factors mentioned earlier poses a challenge, especially as the proliferation of pests and diseases is favored due to the conducive atmosphere created in the greenhouse [41]. Besides the challenges posed by ensuring a balanced micro-climatic condition, the development of pest and disease resistance is also promoted in the greenhouse by other conditions such as the cultivation of only one type of crop (mono-culture). Mono-culture encourages the acclimatization of pests and diseases when one particular crop is cultivated continuously. Likewise, insufficient ventilation and low temperature in the rhizosphere might lead to the multiplication of crop diseases. Other challenges facing greenhouse crop production are the low-cost of agricultural produce compared to the cost of production, which sometimes discourages small and medium-scale farmers [42].

Other non-environmental challenges facing greenhouse production included:

i. High energy consumption in sophisticated greenhouses that operate entirely on air-conditioning, heating, and ventilation systems, lighting, sensors, and other wireless gadgets.

ii. Need for skilled personnel to precisely control the required conditions in the greenhouse, including temperature, humidity, different crop nutrient requirements, and pest management.

iii. In a non-automated greenhouse, the fast and constant depletion of essential nutrients in the liquid media requires careful monitoring and quantification, followed by replenishing the lost nutrients immediately so that additional stress will not be added to the crops.

6.2 Prospects of crop production in greenhouse

The emergence of aquaponics farming systems in the greenhouse indicates a promising prospect in crop production. The aquaponics technique is an intelligent strategy for integrating soilless crop production and fish farming [43]. In this system, the waste generated by the fish after feeding is then broken down into useful nutrients by numerous bacteria species. The nutrients produced in the fish effluents are assimilated by the crops raised in the hydroponics beds. The wastewater becomes clean and then recirculated continuously to the fish tank for reuse with minimal water loss and healthier crops free from chemical fertilizer [44]. Similarly, the incorporation of hydroponics technique into the aquaponics helps in geometrically increasing the production output from a small area of land as crops can be raised in a cascaded manner (either vertically or horizontally) on sterile layers [41], as well as dramatically reducing the spread of diseases that are associated with soil cultivation [38].

Furthermore, the use of an integrated system of wireless sensors, actuators, and robots, eases the stress being faced by farmers as the nutrients levels and other requirements (pH, soil moisture, temperature, humidity) by the crops can be monitored remotely and accurately in real-time, with few physical visits to the greenhouse. Also, the introduction of the Integrated Production and Protection (IPP) method, which is the rational way of combining both the biological and chemical methods of crops disease control with other control measures (good agricultural practice and injection of plant extracts), will in the future enhance crops cultivation in the greenhouse [38].

Soon, a fully automated unmanned greenhouse operated by various energy-saving advanced technologies, including WSN, UAV, and IoT for real-time management of crops, will boost food production and ensure sustainable food safety to consumers.

Author details

Aliyu Idris Muhammad[1*], Abubakar Shitu[1], Umar Abdulbaki Danhassan[2], Muhammad Hilal Kabir[3], Musa Abubakar Tadda[1] and Attanda Muhammed Lawal[1]

1 Department of Agricultural and Environmental Engineering, Faculty of Engineering, Bayero University, Kano, Nigeria

2 Department of Agricultural and Bio-Environmental Engineering, SCA/DAC Ahmadu Bello University, Zaria, Nigeria

3 Department of Agricultural and Bioresource Engineering, Faculty of Engineering and Engineering Technology, Abubakar Tafawa Balewa University, Bauchi, Nigeria

*Address all correspondence to: aimuhammad.age@buk.edu.ng

References

[1] N. Wang and Z. Li, "Wireless sensor networks (WSNs) in the agricultural and food industries," in *Robotics and Automation in the Food Industry: Current and Future Technologies*, 1st ed., D. C. Caldwell, Ed. Woodhead Publishing Limited, 2012, pp. 171-199.

[2] J. A. Aznar-Sánchez, J. F. Velasco-Muñoz, B. López-Felices, and I. M. Román-Sánchez, "An analysis of global research trends on greenhouse technology: Towards a sustainable agriculture," *International Journal of Environmental Research and Public Health*, vol. 17, p. 664, 2020.

[3] N. Tzortzakis, S. Nicola, D. Savvas, and W. Voogt, "Editorial: Soilless cultivation through an intensive crop production scheme. Management strategies, challenges and future directions," *Frontiers in Plant Science*, vol. 11, pp. 10-12, 2020.

[4] R. R. Shamshiri *et al.*, "Advances in greenhouse automation and controlled environment agriculture: A transition to plant factories and urban agriculture," *International Journal of Agricultural and Biological Engineering*, vol. 11, no. 1, pp. 1-22, 2018.

[5] I. V. Pollet, J. G. Pieters, J. Deltour, and R. Verschoore, "Diffusion of radiation transmitted through dry and condensate covered transmitting materials," *Solar Energy Materials and Solar Cells*, vol. 86, no. 2, pp. 177-196, 2005.

[6] R. R. Shamshiri, J. W. Jones, K. R. Thorp, D. Ahmad, H. C. Man, and S. Taheri, "Review of optimum temperature, humidity, and vapour pressure deficit for microclimate evaluation and control in greenhouse cultivation of tomato: A review," *International Agrophysics*, vol. 32, no. 2, pp. 287-302, 2018.

[7] Gorjian, S., Calise, F., Kant, K., Ahamed, M. S., Copertaro, B., Najafi, G., ... & Shamshiri, R. R. (2020). A Review on Opportunities for Implementation of Solar Energy Technologies in Agricultural Greenhouses. *Journal of Cleaner Production*, 124807.

[8] K. Benke and B. Tomkins, "Future food-production systems: Vertical farming and controlled-environment agriculture," *Sustainability: Science, Practice, and Policy*, 2017.

[9] M. A. Akkaş and R. Sokullu, "An IoT-based greenhouse monitoring system with Micaz motes," in *Procedia Computer Science*, 2017.

[10] R. K. Singh, M. Aernouts, M. De Meyer, M. Weyn, and R. Berkvens, "Leveraging LoRaWAN technology for precision agriculture in greenhouses," *Sensors (Switzerland)*. 2020.

[11] C. von Zabeltitz, *Integrated Greenhouse Systems for Mild Climates*. 2011.

[12] T. Rault, A. Bouabdallah, and Y. Challal, "Energy efficiency in wireless sensor networks: A top-down survey," *Computer Networks*. 2014.

[13] D. Li and H. Yang, "State-of-the-art Review for Internet of Things in Agriculture," *Nongye Jixie Xuebao/Transactions of the Chinese Society for Agricultural Machinery*. 2018.

[14] T. Ojha, S. Misra, and N. S. Raghuwanshi, "Wireless sensor networks for agriculture: The state-of-the-art in practice and future challenges," *Computers and Electronics in Agriculture*. 2015.

[15] R. K. Kodali, V. Jain, and S. Karagwal, "IoT based smart greenhouse," in *IEEE Region 10*

Humanitarian Technology Conference 2016, R10-HTC 2016 - Proceedings, 2017.

[16] M. Danita, B. Mathew, N. Shereen, N. Sharon, and J. J. Paul, "IoT Based Automated Greenhouse Monitoring System," in *Proceedings of the 2nd International Conference on Intelligent Computing and Control Systems, ICICCS 2018*, 2019.

[17] W. Dargie, "Dynamic power management in wireless sensor networks: State-of-the-art," *IEEE Sensors Journal*, 2012.

[18] S. Ghiasi, A. Srivastava, X. Yang, and M. Sarrafzadeh, "Optimal energy aware clustering in sensor networks," *Sensors*, 2002.

[19] S. P. Wraight, R. B. Lopes, and M. Faria, "Microbial control of mite and insect pests of greenhouse crops," in *Microbial Control of Insect and Mite Pests: From Theory to Practice*, Elsevier Inc., 2017, pp. 237-252.

[20] S. A. El Arnaouty, A. H. El-Heneidy, A. I. Afifi, I. H. Heikal, and M. N. Kortam, "Comparative study between biological and chemical control programs of certain sweet pepper pests in greenhouses," *Egyptian Journal of Biological Pest Control*, vol. 30, no. 1, pp. 28-34, 2020.

[21] J. B. Jones Jr., *Hydroponics - A practical guide for the soilless grower*, 2nd ed., vol. 53. Florida, USA: CRC Press, 2005.

[22] J. A. Bottomley, "Pest control light," *The Growing Edge*, vol. 12, no. 3, pp. 66-73, 2001.

[23] K. Becker, "Pesticide application equipment," in *Ball Redbook: Greenhouse and Equipment*, 17th ed., C. Beytes, Ed. Batavia, IL.: Ball Publishing, Batavia, IL., 2003, pp. 197-200.

[24] D. Gao, Q. Sun, B. Hu, and S. Zhang, "A framework for agricultural pest and disease monitoring based on internet-of-things and unmanned aerial vehicles," *Sensors (Switzerland)*, vol. 20, no. 5, 2020.

[25] S. Cui, P. Ling, H. Zhu, and H. M. Keener, "Plant pest detection using an artificial nose system: A review," *Sensors (Switzerland)*, vol. 18, no. 2, pp. 1-18, 2018.

[26] FM, "Farming Revolution: The use of sensors in crop pest dectection," 2020. [Online]. Available: https://www.farmmanagement.pro/farming-revolution-the-use-of-sensors-in-crop-pest-detection/. [Accessed: 24-Aug-2019].

[27] Y. Iwasaki, M. Aizawa, C. Yoshida, and M. Takaichi, "Developing a new energy-saving, photosynthesis-promoting environmental control system for greenhouse production based on a heat pump with a heat storage system," *Journal of Agricultural Meteorology*, vol. 69, no. 2, pp. 81-92, 2013.

[28] A. Yano and M. Cossu, "Energy sustainable greenhouse crop cultivation using photovoltaic technologies," *Renewable and Sustainable Energy Reviews*, vol. 109, pp. 116-137, 2019.

[29] J. Bambara and A. K. Athienitis, "Energy and economic analysis for the design of greenhouses with semi-transparent photovoltaic cladding," *Renewable Energy*, vol. 131, pp. 1274-1287, 2019.

[30] N. Yildirim and L. Bilir, "Evaluation of a hybrid system for a nearly zero energy greenhouse," *Energy Conversion and Management*, vol. 148, pp. 1278-1290, 2017.

[31] G. K. Ntinas, M. Neumair, C. D. Tsadilas, and J. Meyer, "Carbon footprint and cumulative energy demand of greenhouse and open-field tomato cultivation systems under

Southern and Central European climatic conditions," *Journal of Cleaner Production*, vol. 142, pp. 3617-3626, 2017.

[32] A. Marucci, A. Gusman, B. Pagniello, and A. Cappuccini, "Limits and prospects of photovoltaic covers in Mediterranean greenhouses," *Journal of Agricultural Engineering*, vol. 43, no. 4, p. 1, 2013.

[33] L. Mariani, G. Cola, R. Bulgari, A. Ferrante, and L. Martinetti, "Space and time variability of heating requirements for greenhouse tomato production in the Euro-Mediterranean area," *Science of the Total Environment*, vol. 562, pp. 834-844, 2016.

[34] N. S. Mattson and J. E. Erwin, "The impact of photoperiod and irradiance on flowering of several herbaceous ornamentals," *Scientia Horticulturae*, vol. 104, no. 3, pp. 275-292, 2005.

[35] R. Shamshiri, P. van Beveren, H. Che Man, and A. J. Zakaria, "Dynamic assessment of air temperature for tomato (*Lycopersicon esculentum* mill) cultivation in a naturally ventilated net-screen greenhouse under tropical lowlands climate," *Journal of Agricultural Science and Technology*, 2017.

[36] Shamshiri, R. R., Bojic, I., van Henten, E., Balasundram, S. K., Dworak, V., Sultan, M., & Weltzien, C. (2020). Model-based evaluation of greenhouse microclimate using IoT-Sensor data fusion for energy efficient crop production. Journal of Cleaner Production, 121303.

[37] D. Ma, N. Carpenter, H. Maki, T. U. Rehman, M. R. Tuinstra, and J. Jin, "Greenhouse environment modeling and simulation for microclimate control," *Computers and Electronics in Agriculture*, 2019.

[38] W. Baudoin *et al.*, *Good Agricultural Practices for Greenhouse Vegetable Crops: Principles for Mediterranean Climate Areas*. Rome, 2013.

[39] I. Attar, N. Naili, N. Khalifa, M. Hazami, M. Lazaar, and A. Farhat, "Experimental study of an air conditioning system to control a greenhouse microclimate," *Energy Conversion and Management*, vol. 79, pp. 543-553, 2014.

[40] S. Ghani *et al.*, "Design challenges of agricultural greenhouses in hot and arid environments – A review," *Engineering in Agriculture, Environment and Food*, vol. 12, no. 1, pp. 48-70, 2019.

[41] Auroras, "Greenhouse Cultivation: problems and solutions," *Agriculture*, 2019.

[42] P. Borse, "Challenges Facing Greenhouse Agriculture." Loksatta Team, India, 2020.

[43] C. Maucieri *et al.*, "Life cycle assessment of a micro aquaponic system for educational purposes built using recovered material," *Journal of Cleaner Production*, vol. 172, pp. 3119-3127, 2018.

[44] D. C. Love, M. S. Uhl, and L. Genello, "Energy and water use of a small-scale raft aquaponics system in Baltimore, Maryland, United States," *Aquacultural Engineering*, vol. 68, pp. 19-27, Sep. 2015.

Chapter 6

Greenhouse Crop Simulation Models and Microclimate Control Systems, A Review

Seyed Moin-E-Ddin Rezvani, Redmond R. Shamshiri, Ibrahim A. Hameed, Hamid Zare Abyane, Mohsen Godarzi, Davood Momeni and Siva K. Balasundram

Abstract

A greenhouse is a complex environment in which various biological and non-biological phenomena occur. For simulation and prediction of the climate and plant growth changes in the greenhouse are necessary to provide mathematical models. The dynamic greenhouse climate models are classified in mechanistic and black-box models (ARX). Climatic models are mainly obtained using energy balance or computational fluid dynamics. In the energy balance models, the greenhouse climatic variables are considered uniformity and homogeneity, but in the computational fluid dynamics, the heterogeneity of the greenhouse environment can be shown by 3D simulation. Crop growth simulation models are quantitative tools based on scientific principles and mathematical relationships that can evaluate the different effects of climate, soil, water, and crop management factors on crop growth and development. In this chapter, with a review of the basics of climate models in greenhouses, the results and application of some climate dynamics models based on the energy balance as well as simulations performed with computational fluid dynamics are reviewed. A review of greenhouse growth models and functional–structural plant models (FSPM) was also conducted.

Keywords: Crop, Computational Fluid Dynamics, Energy balance, Functional-structural plant, Greenhouse Climate, Growth model

1. Introduction

A greenhouse is a complex environment in which a variety of physical and biological phenomena occur that affect each other. To understand the complexities of the crop responses to its environment and management practices as well as the dynamics of the greenhouse environment in response to external conditions, greenhouse characteristics, and management, the researchers have been developing models of greenhouse environments and crops [1]. The microclimate and the plant are two subsystems in the greenhouse that exchange matter and energy. Changes in temperature, humidity, light, and concentration of carbon dioxide in the greenhouse affect the crop, and the crop also affects the microclimate by producing latent and sensible heat (transpiration) and photosynthesis. Due to the importance of greenhouse climatic conditions and their effect on quantitative and qualitative crop yield, the use of mathematical models to study and microclimate simulation of the greenhouse is necessary. Greenhouse climate simulation models are used to describe the interactions between greenhouse plant processes (photosynthesis and transpiration) and greenhouse climate including structure shape, cover characteristics, climate control equipment, and surrounding weather conditions [2]. The enormous variety of boundary conditions and design elements makes analyzing greenhouse climate a complex task. Simulation tools are an indispensable support for greenhouse climate studies because they make it possible to take all of these characteristics into account [3]. Therefore, such a model can serve for the optimization of greenhouse design, climate control, and crop management [2]. Plant growth is a complex phenomenon that depends on soil, plant, climate, and their interactions [4]. The crop growth model is an essential part of optimizing crop management [5]. The functional-structural plant models (FSPM) community is developing models to understand the biological processes involved in plant performance and growth. Recently, due to the increasing computing power of computers, three-dimensional models of plants are used to understand the biological processes involved in crop yield and growth and to understand the interaction. The three-dimensional structure of the vegetation with the surrounding environment has been considered. The development of climate models and crop growth models with the help of the Internet of Things and cloud computing has led to attention being paid to digital twin greenhouses.

2. Mathematical models

A summary of reality is called a model. In other words, the abstract or physical representation of an object or system (from a particular point of view) is a model. Modeling helps researchers systematically analyze various scenarios in the greenhouse and predict their behavior. Models come in many forms (such as physical models, mathematical models, and statistical models) and have a variety of vital applications in all areas of science and technology. A mathematical model is a description of a system using mathematical language and its theorems and symbols. The process of creating and selecting models is called modeling. Mathematical methods and complexity in solving these models become an art that only a person who is deeply involved in it can use successfully. Modeling or mathematical modeling is the attempt to develop a mathematical model for a given system. Lumped mechanistic model, static model, steady-state model, black-box model, dynamic simulation model, mechanistic model, stochastic model, heuristic model, descriptive model, explanatory model, state variable model, and distributed fluid dynamics model are examples of mathematical modeling to solve world problems, although modeling is not limited to the above. Such a model is believed to be a simplified representation of a system to serve particular purposes. The modeling purposes are (a) knowledge integration, (b) testing hypotheses, (c) estimating the effect of conditions beyond the range of experimental data, (d) showing knowledge gaps and determine the research pathway, and (e) helping make practical decisions (input of resources, climate control in greenhouses, planning of processes) [6].

3. Climate modeling in greenhouse

The dynamic models of greenhouse climate are classified in mechanistic and black-box models (ARX) [7]. Mechanistic models describe the system it is simulating based on knowledge of the processes that are taking place [8]. Whereas black-box models are more used for applications that involve control, optimization, and design of the greenhouse system [7]. Mechanistic models are based on physical Equations [9] and give the opportunity to be used for intelligent decision support on climate control actions [10]. They enable a quantitative approach of the greenhouse system as transparent mechanistic models, allowing for optimization algorithms to find an optimal control, and they are physically interpretable. The black box model allows statistical description based on the outputs, given inputs on a limited range [9]. Black box models do not suffer from the need to determine the value of each individual parameters. The model only uses data obtained from direct measurements and is considered an empirical approach. So, this system also provides a description of the climate of a greenhouse [9]. These models can be used to estimate the inside environment changes and they can be very helpful for climate control purposes [11]. One of the first mechanistic dynamic models of greenhouse climate was developed by Bot [12] and the first model of the greenhouse climate with optimal control purposes was proposed by Van Henten [13]. Taki et al. [14], using a multilayer perceptron neural network (MLP) model to predict greenhouse temperature, showed that the MLP model can predict the greenhouse climate with a lower Mean Absolute Percentage Error (MAPE) than a dynamic model.

The computational fluid dynamics (CFD) method is the most commonly used for simulating situations where airflow is an important component. In the greenhouse climate study, indoor environmental conditions depend on ventilation efficiency. As a result, indoor greenhouse variables such as temperature, pollution, and humidity are controlled by airflow patterns. Therefore, understanding the principles of air movement is essential to study the greenhouse environment. Computational fluid dynamics (CFD) is a powerful tool that makes it possible to predict the distribution of the climatic variables inside a greenhouse. This numerical device also makes it possible to test different scenarios without the need of experimental approach. The effect of the crop as a plant model on the climatic distribution in a greenhouse has been studied by considering the cultivation of lettuce [15] tomato [16–18]; rose [19] and begonia [20]. In these studies, the Jarvis model [21] has been used to simulate stomatal resistance, in which microclimate and transpiration rate distribution in the greenhouse has been extensively investigated in the past through CFD tools. The main disadvantage of computational fluid dynamics is the high computational costs required. These limitations limit the simulation efficiency to short periods and to identify a limited set of possible scenarios.

3.1 Componential fluid dynamic (CFD)

Computational Fluid Dynamics (CFD) is a numerical solution of the energy balance of a controlled volume that provides the ability to predict the distribution of climatic variables within the greenhouse. Over the last decade, modeling of crop interaction with the climate indoor the greenhouse has been studied. The effect of the crop on the pressure drop inside the flow causes a momentum sink. The crop is assumed as a porous medium in the Darcye Forchheimer equation to estimate the source term. It is however generally assumed that pressure forces contribute the major portion of total canopy drag and, consequently, that the viscous resistance of the crop may be neglected. The crop acts as a sink or source of heat or water vapor, which exchanges matter and energy with the surrounding environment in the form of sensible heat and latent heat flux. The sub-model of the crop was explained by Boulard and Wang [15] and then used and improved by other authors.

3.1.1 Fundamental equations

Computational fluid dynamics techniques manage the values of dependent variables as initial unknowns in a limited number of points, and then a set of algebraic equations derived from the basic equations used in the domain are solved by predefined algorithms. The three basic physical principles identified by the well-known Navier–Stokes equations are mass, momentum, and energy conservation. For an uncompressible fluid, the three-dimensional conservation equations describing the transport phenomena for steady flows in free convection are of the general form:

$$\frac{\partial(\rho\phi)}{\partial t} + \nabla.(\partial\overline{u}\phi) = \nabla.(\Gamma\nabla\phi) + S_\phi \tag{1}$$

where ρ is density (kg m^{-3}), t is time, ∇ is divergence operator, and φ represents the concentration of the dimensionless transported quantity, namely momentum, mass (air and water vapor mass fraction) and energy, and u (m s^{-1}) is the components of the velocity vector. Γ is the diffusion coefficient (m^2 s^{-1}) and S_ϕ is the source term that indicates changes in the amount of matter in the transfer. The diffusion sentence is affected by a coefficient Γ which can be the mass diffusion coefficient (D), the momentum diffusion coefficient (μ), and the energy diffusion coefficient (k). Consequently, turbulence models must be introduced in the Reynolds equations written to separate the mean flow from its fluctuating components. One of the most widely used closure procedures is the k-ε model which introduces two new phenomenological variables: the turbulent kinetic energy k, and its dissipation rate.

3.1.2 Radiative submodel

The discrete ordinate (DO) model is used to calculate the radiant heat transfer caused by the sun rays on semitransparent walls and borders. Discrete ordinate model can solve the problem,using the gray range model and assuming gray or non-gray radiation. In the discrete ordinate radiation model, the separate directions of the radiation transfer equation in the S direction are written as a field equation. The discrete ordinate (DO) equation is [22]:

$$\nabla.\left(I_\lambda\left(\vec{r},\vec{S}\right)\vec{S}\right) + (\alpha_\lambda + \sigma_s)I_\lambda\left(\vec{r},\vec{S}\right) = \alpha_\lambda n^2 I_{b\lambda} + \frac{\sigma_s}{4\pi}\int_0^{4\pi} I_\lambda\left(\vec{r},\vec{S}\right)\phi\left(\vec{S},\vec{S}\right)d\Omega' \tag{2}$$

where $\vec{r}$ is the position vector, $\vec{S}$ is the vector for radiation direction, I_λ is the radiation intensity (W m^{-2} src^{-1}) which depends on position $\left(\vec{r}\right)$ and direction $\left(\vec{s}\right)$, α_λ is the absorption coefficient (m^{-1}), σ_s is the scattering coefficient (m^{-1}), ϕ is the phase function, Ω' is the solid angle (deg), and n is the refractive index.

3.1.3 Crop sub-model

The sink of momentum due to the drag effect of the crop, is symbolized by the source term S_ϕ of the Navier–Stokes equation. This drag force may be expressed by the unit volume of the cover by the commonly used formula [23]:

$$S_\phi = \frac{\partial p}{\partial x} = -\rho L C_D u^2 \tag{3}$$

where u is the air speed, L the leaf area density ($m^2\ m^{-3}$) and C_D a drag coefficient. For a mature greenhouse tomato crop C_D is 0.32.

3.2 Review of CFD studies in greenhouse

Kacira et al. [24] investigated the effect of wind speed, side vents, and the number of spans on natural ventilation of a multi-spans greenhouse by numerical simulation in a CFD software. The results showed that the maximum amount of greenhouse ventilation was achieved by the simultaneous use of side and roof vents. The ratio of ventilation to the ground area of the greenhouse was 9.6%, which was lower than the recommended ratio of 15 to 25%. The results showed a significant decrease in the amount of ventilation with increasing the number of openings. Baeza et al. [25] considered the cooling of a greenhouse with natural ventilation to be conditional on the creation of a suitable and sufficient combination of air exchange from the roof and side vents to remove the excess heat of the sensible heat by moving the air through the vegetation. The CFD simulation results showed the most of the exchanged air was at the top of the canopy and warm harmful areas were created inside the canopy due to the slow movement of air. Flores-Velázquez [26] in a study using CFD showed that the wind pattern inside the greenhouse was greatly influenced by the wind speed outside the greenhouse and the number of greenhouse spans. In the case of greenhouses with 3 or 4 spans, ventilation was independent of the roof vents, but in greenhouses with five or more spans, side ventilation prevailed over roof ventilation. Flores-Velazquez et al. [27] in a study investigated the temperature exchange and distribution in a greenhouse with natural and mechanical ventilation systems with 30 and 100% openness of roof vents in four greenhouse lengths of 28, 50, 75, and 100 m. The results showed a strong linear relationship between temperature slope and greenhouse length. Simultaneous use of roof ventilation and fan compared to mechanical ventilation alone improved the air exchange rate (22%) and the uniformity of the greenhouse climate. With increasing the length of greenhouses, the advantage of natural ventilation over mechanical ventilation was greater. Increasing the capacity of the fans generally reduced the temperature, but the effect was less severe in openness greenhouse roof vents. Roy et al. [28] simulated the temperature and humidity distribution in a semi-enclosed 960 m^2 glasshouse with a tomato crop using CFD. Radiation exchange simulation was modeled with a separate directional radiation model (DO) and tangible, latent and radiant heat transfer along with product activity (aperture resistance) and water vapor transfer in vegetation with a porous and semi-transparent medium. To limit the computation time and the size of the grid, the geometric domain was limited to the greenhouse walls. The simulation values and experimental data were generally in good agreement, however, a disagreement between the two was evident for the concentration of water vapor during the opening period of the valves. These differences indicated the impossibility of obtaining accurate simulation values with a limited amplitude when the skylights are open. Simulation of temperature and water vapor concentration patterns inside the greenhouse showed that doubling the airflow rate leads to a significant change in the climate distribution inside the greenhouse. Higher values of airflow rate do not further change these parameters. Flores-Velázquez et al. [29] used CFD modeling in tomato-growing greenhouses in a study aimed at proposing alternatives to environment management, estimating energy costs, and the economic costs of using fans. Results showed that in areas with mild summers, the use of mechanical and natural ventilation together is a suitable alternative to reduce temperature and energy costs. The use of combined ventilation due to high temperature in the hours of maximum radiation, while reducing problems, does not affect production costs.

Molina-Aiz et al. [30] investigated the effects of four different greenhouse vents arrangements and two outside wind speeds on ventilation and temperature distribution in uncultivated greenhouses in the Almera region of Spain. The results showed that the ventilation vents arrangement was affected by airflow, ventilation intensity, and air temperature distribution in the greenhouse. For different settings, the wind direction perpendicular to the openings and outside wind speed 5 m s^{-1}, calculated airflow from 70.1 to 134.9 m^3 s^{-1} and for wind speed 1 m s^{-1} from 21.0 to 43.3 m^3 s^{-1} were variable. Bournet et al. [31] studied the combination of the side and roof vents with consideration of ventilation, airflow patterns, and temperature distribution in a four-span greenhouse (2600 m^2) equipped with continuous roof vents and benches supporting ornamental crops. The results showed that the arrangement with side and roof vents led to a maximum flow rate of 12.3 times air exchange per hour with a wind speed of 0.15 m s^{-1}.

Fatnassi et al. [19] investigated the effect of insect screens on airflow and climatic conditions of multi-span 1000 m^2 square greenhouses by CFD. The main results showed that the increase in temperature and humidity due to the use of insect screens can be corrected by the simple arrangement of the system, such as the intelligent selection of roof vents and the use of additional side vents. Bartzanas et al. [32] simulated the effect of insect screens in a tunnel greenhouse with a tomato crop and showed that reducing the porosity of the insect screens led to a gradual increase in temperature and humidity and a decrease in air velocity. Majdobi et al. [33] in another simulation in the same greenhouse showed that insect screens reduce the ventilation rate by 46%.

Chen et al. [34] adapted the CFD model to simulate the distribution of velocity and air temperature in greenhouses with fan and pad cooling systems in summer. The CFD simulations showed that when the crop canopy height is between 2 to 3 m, the fan and pad height options from the ground level of 0.6 and 1.4 m, respectively, and the fan and pad height of both 1.4 m were appropriate.

Fidaros et al. [35] investigated the simultaneous effect of solar radiation distribution and, ventilation in a tunnel tomato greenhouse at different summer days hours. Simulations showed the effect of the angle of incidence of the incoming radiation radius on the distribution of solar radiation inside the greenhouse. Another result was the predominance of forced convection due to mechanical ventilation. Baxevanou et al. [36] simulated the effect of solar radiation distribution in a tunnel greenhouse in two dimensions with discrete ordinate (DO) model according to the thickness, optical and thermal properties of the cover. The results showed that the greenhouse cover with high absorption of solar radiation disrupts natural ventilation, increases the air temperature inside the greenhouse due to the phenomenon of convection and the development of secondary recirculation. At the same time, high absorption reduces photosynthetically active radiation (PAR). Ortiz-Vazquez et al. [37] reported that the main problem of large commercial greenhouses is inadequate ventilation and increased production costs due to the use of air conditioning. They studied the environmental conditions of a large commercial greenhouse to optimize the design using CFD and showed that the large geometry of the greenhouse and the height of the greenhouse cover determine the impact of incoming radiation inside the greenhouse and thus the environmental conditions.

Ali et al. [23] a special sub-model to simulate the distribution of transpiration and climate around potted plants in water-restricted greenhouses in the form of a 2D transient CFD model with user-defined functions to match product interaction by developed the climate inside the greenhouse. The crop was considered as a porous medium, and special source terms for transpiration and sensible heat transfers were added. The simulation results showed the model's ability to accurately predict transpiration, air temperature, leaf, and indoor air humidity in irrigation regimes. Experiments showed that water supply can be reduced without a significant effect on transpiration rate and thus plant growth potential up to 20%. Piscia et al. [3] proposed a CFD model for climatic simulation and night condensation in a plastic-covered four-span greenhouse. The results, while displaying the importance of radiation heat loss losses, showed that the greenhouse roof is the coolest surface for condensation of water vapor produced by the crop.

Boulard et al. [38] developed a CFD model for predicting the distribution of temperature, water vapor, and carbon dioxide in a semi-enclosed glass greenhouse equipped with an air conditioning system. Sensible and latent heat fluxes in the crop rows were added to the main model through the radiation model of discrete ordinate (DO) and changes in carbon dioxide concentration through the photosynthesis model. The simulated values of temperature, humidity and carbon dioxide concentration had a good agreement with the measured values. The simulation results for investigating the vertical distribution of temperature and humidity for two leaf area density of 2.95 m^{-1} and 5.9 m^{-1} showed that long and dense vegetation intensifies the cooling of indoor air as well as increases temperature layering.

Rezvani [39] investigated mechanical ventilation and pad-fan to improve the climatic conditions in an asymmetry commercial greenhouse with an area of 4333 m^2 in June. Ansys Fluent 16 software was used to prepare the computational fluid dynamics model, and to calculate the rate of penetration and absorption of radiation by vegetation, vapor pressure deficit, leaf temperature, and transpiration was coded using the user-defined function (UDF). The results showed that natural ventilation could not improve the climatic conditions inside the greenhouse and the use of climate control equipment is necessary. Mechanical ventilation also causes more uniformity and homogeneity of the greenhouse climate, but depending on the climatic conditions around the greenhouse may not be able to improve the greenhouse climate. The computational fluid dynamics model simulated the real conditions properly and showed that the addition of a pad-fan in the greenhouse in June could reduce the temperature from 38.0 to 22.7, increase the relative humidity from

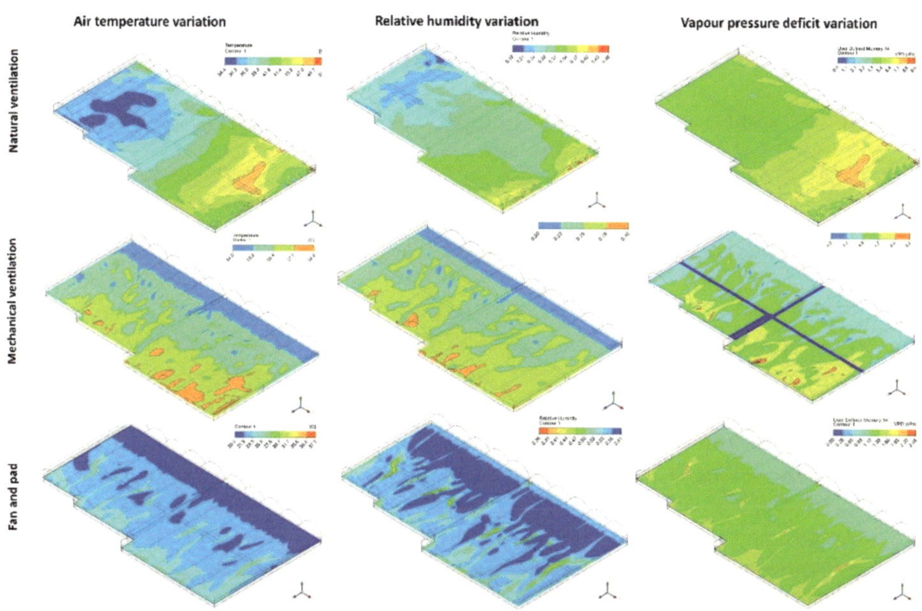

Figure 1.
Simulation of temperature, relative humidity and vapor pressure deficit in natural ventilation, mechanical ventilation, and cooling system (fan and pad) in a commercial greenhouse. Simulation time is 14:00 on 10 June 2018.

29.5% to 55.5% and reduction of vapor pressure deficit from 4.51 to 1.26 kPa and optimize the environmental conditions for the product (**Figure 1**). The results showed that in large commercial greenhouses, it is better to prevent non-uniformity due to structural asymmetry by constructing a greenhouse structure symmetrically.

4. Energy and mass balance

In the energy balance model, the greenhouse is considered as a "perfectly stirred tank" assuming the uniformity and homogeneity of the greenhouse variables such as temperature and humidity [40]. This assumption causes the energy balance model computationally is done fast and explicit, but on the other hand, it is a source for certain limitations. Calculations in the energy balance model require some priori and empirical information of different coefficients such as ventilation intensity and heat transfer coefficient. Energy balance simulation is based on the analysis of heat and mass balance equations used throughout the greenhouse system [41].

Energy and mass balance equations are used to estimate temperature, absolute humidity and CO_2 in a greenhouse [42, 43]:

$$\frac{dT}{dt} = \frac{1}{C_{cap}}\left(Q_{sun} - Q_{cov} - Q_{trans} + Q_{lamp} - Q_{vent} + Q_{he,heat} - Q_{he,cool} + Q_{pipe}\right)\ (°C\ s^{-1}) \quad (4)$$

$$\frac{d\chi_{air}}{dt} = \frac{1}{h}\left(\phi_{trans} - \phi_{cov} - \phi_{he,cool} - \phi_{vent}\right)\left(g\ m^{-3} s^{-1}\right) \quad (5)$$

$$\frac{dCO_{2,air}}{dt} = \frac{1}{h}\left(\phi_{c.inj} - \phi_{c,ass} - \phi_{c,vent}\right)\left(g\ m^{-3} s^{-1}\right) \quad (6)$$

where h is the average height of the greenhouse, incoming radiation Q_{sun}, heat losses through the cover Q_{cov}, transpiration by the crop Q_{trans}, artificial lighting Q_{lamp}, natural ventilation Q_{vent}, heating $Q_{he,heat}$ and cooling $Q_{he,cool}$ with the heat exchangers, and heating by the pipe rail system Q_{pipe} ($W\ m^{-2}$). The vapour balance is influenced by crop transpiration ϕ_{trans}, condensation on the cover ϕ_{cov}, condensation in the heat exchangers due to cooling $\phi_{he,cool}$, and vapour exchange with outdoor air by natural ventilation ϕ_{vent} ($g\ m^{-2}\ s^{-1}$) (**Figure 2**). $\phi_{c,inj}$ is the injection of pure industrial CO_2 to the greenhouse, $\phi_{c,ass}$ is the assimilation of CO_2 by the crop, and $\phi_{c,vent}$ is the CO_2 exchange with outside.

4.1 Review of dynamic climate models studies in greenhouse

Kindelan [44] used a dynamic model to simulate the indoor environmental conditions of the greenhouse. The external climatic variables used in the initial simulation were: constant wind speed, constant relative humidity, solar radiation, and air temperature. In order to simulate the internal environment by the energy balance method, the system is divided into four elements; soil, plant, internal air and cover, modelling the heat and mass fluxes between these elements. The dynamic model was used to predict temperature, humidity, heat flow inside the greenhouse using ventilation conditions, night heating, and vegetation percentage.

A dynamic mechanical model of the greenhouse climate developed by Bot [12] includes four temperature variables: greenhouse cover, air, crop canopy, and four layers of soil, as well as air and soil moisture. The physical processes in energy balance were transient, convection, conduction, ventilation, and radiation. Mass balance, on the other hand, is considered convection, ventilation, transpiration, and

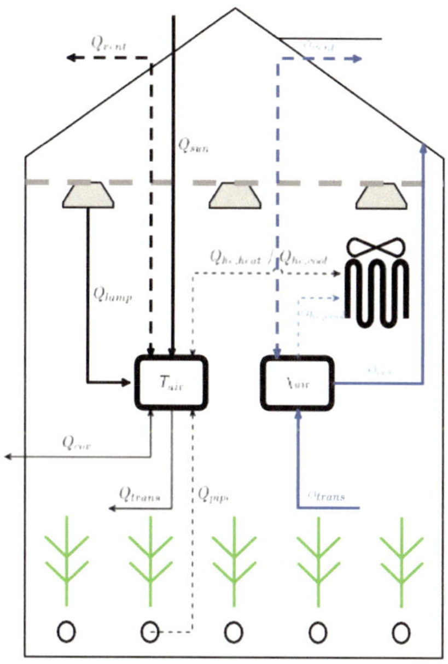

Figure 2.
Display Tair and χair climatic variables, and related energy fluxes Q and vapor fluxes ϕ (source: [42]).

condensation. The input variables were temperature, humidity, sunlight, and wind speed outside the greenhouse. In a more detailed model, the concentration of carbon dioxide in the air was considered as another variable [45] in the greenhouse. Therefore, the physical and biological processes in the more accurate model were CO_2 injection, product photosynthesis, and product respiration rate. In the latter model, the temperature of the heating pipes was also in the energy balance.

The greenhouse process (KASPRO) model is constructed from modules describing the physics of mass and energy transport in the greenhouse enclosure, and a large number of modules that simulate the customary greenhouse climate controllers [46]. The state variables included air temperature, carbon dioxide concentration and humidity (water vapor or partial vapor pressure). In the case of the temperature, unsteady-state balances were carried out at the greenhouse cover, the air above a thermal screen, air below a thermal screen, on the crop canopy, in the floor and in six soil layers. For carbon dioxide and air humidity, mass balances were performed above and below a thermal screen. The standard heat exchange theory is used for the convective heat exchange between all the surfaces calculated. The KASPRO climate model can also be used to control heating, ventilation, dehumidification, humidification, shading, artificial light and carbon dioxide supply. The model can also be used to characterize the performance of the boiler, short-term and seasonal heat storage facilities, simultaneous production of heat and electricity and heat pumps. The input variables were temperature of the sky, temperature outside the greenhouse, temperature of a deep soil layer, temperature of upper and lower heating pipe, external vapor pressure, carbon dioxide concentration and wind speed outside the greenhouse. Control variables were CO_2 supply, window aperture, thermal screens and artificial lighting. The physical processes involved in the mass and heat unsteady-state balances included radiative heat exchange, convective exchange, ventilation, condensation, transpiration and conduction. This model was developed for glasshouses [46].

Vanthoor [47] developed a model to study the effects of outdoor climate and greenhouse design on the indoor greenhouse climate. To use the greenhouse design method, which focused on optimizing a set of design elements, this model had to meet the following three: 1) Predicting the temperature, vapor pressure, and CO_2 concentration greenhouse indoor air, for different greenhouse designs and climatic conditions, 2) Ability to consider greenhouse construction parameters and climate conditioning equipment, 3) Possibility to make it combine it with tomato yield model. The dynamic model was approved for four different greenhouse designs under three climatic conditions: a temperate marine climate, a Mediterranean climate and a semi-arid climate.

Mobtaker et al. [48] investigated six greenhouse forms with north–south and east–west orientations in terms of energy consumption in the climate of Tabriz, Iran. The minimal extra thermal energy required to maintain suitable temperature conditions for plant growth was observed in a single-span greenhouse in an east–west direction with a north brick wall. The results showed that the northern brick wall could reduce greenhouse heating demand by 31.7%.

Using a single-span semi-solar greenhouse for experimental research designed and built by Mohammadi et al. [11], Mobtakar et al. [48] developed a dynamic model for predicting indoor air temperature in the greenhouse. The results showed that the predicted and measured data are consistent.

Taki et al. [14] investigated a dynamic climate model in a semi-solar greenhouse was designed and constructed at the North-West of Iran in Azerbaijan Province. Crop, soil, cover and thermal screen temperature, and air temperature below and above the screen were measurement. Then the temperature in different parts of the greenhouse was estimated by a dynamic heat and mass transfer model with initial values and considering the evapotranspiration of the crop. It was reported that the predicted and experimental data were in good agreement. Yildiz and Stombaugh [49] developed a dynamic simulation model for predicting climate in the greenhouse as a function of dynamic environmental factors. The model can consider the effects of location, time of the year, orientation, single and double polyethylene glazing, conventional and heat pump heating and cooling systems, open and confined greenhouse systems, CO_2 enrichment, variable shading, and the use of night curtains. The greenhouse heating and cooling systems were a conventional gas furnace and evaporative cooling, respectively. The model was able to simulate the temporal and vertical distribution temperatures of air, leaf, floor, and cover. Also, the model simulated relative humidity, CO_2, photosynthetic active radiation, respiration, transpiration, energy, and CO_2 utilization. Comparison of measured and predicted results showed that the simulated and predicted parameters are in good agreement.

Salazar-Moreno et al. [50] used a dynamic energy balance model to predict the temperature in a 120 m^2 greenhouse with polyethylene cover, natural ventilation, and tomato cultivation in central Mexico. The model considered plant transpiration, ventilation, condensation inside the greenhouse, outdoor climatic conditions, crop characteristics (leaf area index, stomatal and aerodynamic resistance), cover properties and greenhouse characteristics. The results showed that the mean absolute error (MAE), root mean square error (RMSE), and model efficiency (EF) were 1.86°C, 2.256, and 0.657, respectively. After calibrating, the model efficiency increased to 33.84% and received 0.89. Although the predicted transpiration was not close to the values in the sources, the model was effective as a tool to analyze the temperature behavior in the greenhouse.

Joudi and Farhan [51] developed a dynamic model that considered the exchange of soil surface heat with greenhouse air to more accurately predict indoor temperatures. The input parameters of the model were meteorological conditions and thermal properties of the indoor air, cover, and soil of the greenhouse. The results showed that the simulated and experimental data were in good agreement.

5. Crop growth model

Crop growth model is an essential part of the optimization for cultivation management [5]. Crop growth simulation models are quantitative tools based on scientific principles and mathematical relationships that can evaluate the different effects of climate, soil, water, and crop management factors on crop growth and development. Based on the advances made, today the approach of using computer software to manage agricultural production systems is considered a powerful tool. Crop production management (irrigation, fertilization, pest and disease control), climate change, climate fluctuations, yield forecast, environmental pollution, sustainable agriculture, and many other aspects are studied with the above approach.

The first crop growth models were built for open field crops, and greenhouse crops models were developed decades later. There is little difference between farm and greenhouse crop growth models [52].

The main reforms that were needed included the following: modified of radiation conditions due to greenhouse cover, use of complementary lighting and screens (and rapid changes in radiation), extreme climatic conditions in winter and summer, a more detailed description of the effects of temperature on crop performance, effect of CO_2 concentration, and the crucial role of maintenance respiration in winter cultivation [52]. Process-based models are mostly used for crop model development. In process-based models, the rates of growth and development are derived from basic principles in heat and mass transfer and plant physiology. Common processes for plants are photosynthesis, respiration, growth and development. According to the model application, the stages of plant flowering and fruiting may be added to it [53].

Growth models consist of two categories: descriptive and explanatory models. A descriptive model, based on existing theoretical knowledge and practical experience, determines the relation between the research factors using regression analysis of mass crop data. The explanatory model describes the relationship between environmental factors, crop management and crop growth, morphological growth and yield formation process based on the principle of dynamics [5].

5.1 Crop growth models in greenhouse

There is a wide range of explanatory models for greenhouse crops, with several more prominent models including TOMGRO and HORTISIM for general crops [53–55]. Some of the well-known simulation models for tomato plants [56] include TOMSIM and SUSROS87 [57], TOMPOUSSE [58], TOMGRO [1, 59]. There is seen to be a common weakness with these models in that their parameters are specific for the climate condition and greenhouse design they were derived from. In addition, the complexity of the interactions between the greenhouse elements and the crop itself, makes it often impossible to correctly predict microclimate effects on the final yield with the same model parameters. In Jones et al. [1] study, it was claimed that the simplified TOMGRO model is possible to use for different climate conditions with the same parameters derived from their experiment. The first version of TOMGRO [59] and the third version, respectively had 69 and 574 state variables for the simulation of tomato growth on the basis of three inputs measured inside the greenhouse environment: the photosynthetically active radiation in [μmol m^2 s^{-1}], air temperature [°C] and CO_2 concentration CA [ppm]. A simplified version of TOMGRO [1] was developed with the objective of providing a practical application which only had five steady-state variables: (i) node number for the main stem, (ii) leaf Area Index, (iii) total plant dry weight (WT), (iv) total fruit dry weight (WF), and (v) mature fruit dry weight (WM). Some of the studies related to the evaluation and adaptation of TOMGRO model to specific climate conditions and cultural practices, a few can be found in [5, 60, 61]. It should be noted that the simplified TOMGRO model only takes into account the effect of air temperature and light condition, while other important variables such as CO_2 concentration were not included in this version.

A model-based method to design greenhouses for a broad range of climatic and economic conditions, a tomato yield model developed by Vanthoor et al. [62] describes the effects of greenhouse climate on yield. The tomato yield was simulated for various light conditions and concentrations of CO_2 for optimal and non-optimal temperatures in the Netherlands and southern Spain. The model simulated the effects of extremely low as well as average high temperatures on the yield and harvest time of the first fruit with acceptable accuracy [62]. Lin et al. [5] stating that today the models for predicting the performance of the greenhouse crop have their specific application conditions, which may not ensure the accuracy of the results if the greenhouse environment changes. To solve this problem, they studied two widely used tomato growth models TOMGRO and Vanthoor, and then proposed an integrated model. Results showed compared with TOMGRO and Vanthoor models, the output of the integrated model was more reasonable and universal, and the model output was closer to the actual value.

5.2 Functional–structural plant (FSP) modeling

The need to integrate expanding knowledge into the plant sciences has led to the development of advanced modeling approaches, such as functional-structural plant (FSP) modeling which is the result of cross-fertilization between the domains of plant science, computer science, and mathematics [63]. These models provide an opportunity for computational botany to address issues in complex plant systems that cannot be fully explained by experimental approaches alone. FSP modeling is now a well-established approach that has been perfected over the years. FSP models simulate growth and morphology of individual plants and their interaction with the environment, from which the complex properties of the plant community emerge [64]. Investigated the distribution of light interception in a canopy, optimal pruning strategies in orchards, and grass branching about plant population density leading to take into account plant architecture and its development as an integral component [64]. Functional–structural plant (FSP) models have been used widely for over two decades to understand the complex interactions between plant architecture and underlying processes driving plant growth [65]. Functional–structural plant models (FSPMs) were initiated after the concepts of plant architecture became widely acknowledged in botany and in parallel with development of the computational power offered by personal computers [65]. FSP modeling makes it possible to simulate the three-dimensional structure of each plant individually over time. The three-dimensional structure can consider the retention and scattering of light on the leaf surface as a function of leaf size, angle and optical properties. The obtained information can be used to characterize photosynthesis, photomorphogenesis, and overall plant growth and development. Therefore, by considering phenotypic variability between individuals and plastic responses to environmental conditions, as well as by changing plant architecture such as pruning or herbivore, FSP modeling can be used to understand the behavior of a single plant to the performance of the entire canopy [64]. Conceptual diagram of functional-structural plant modeling, which can be formed for scaling from gene to community integration level (**Figure 3**). FSP models typically simulate the three-dimensional structure of plants as a result of individual plant growth, driven by plant physiological processes,

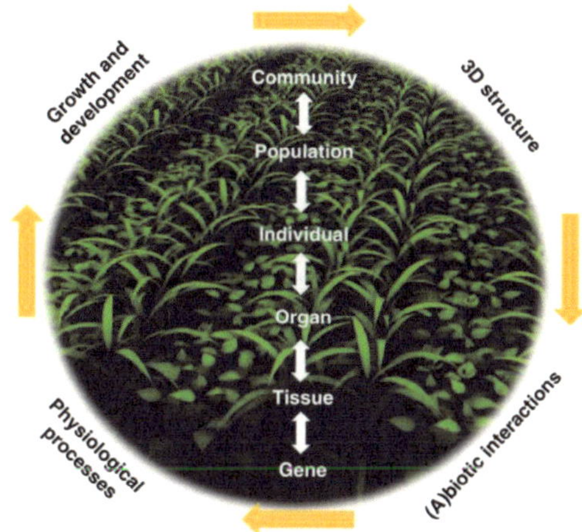

Figure 3.
Conceptual diagram of functional–structural plant (FSP) modeling (source: [63]).

which in turn are influenced by biological and non-biological factors of vegetation (Light, temperature, fungi, insects, etc.). In turn, the distribution of these factors in the canopy is determined by the three-dimensional structure of the plants.

6. Greenhouse climate and crop models combination

Vanthoor [47, 62, 66, 67] developed a greenhouse environment system model that includes a greenhouse microclimate model, a greenhouse growth model, and an economic model.

An FSP combined with a climate model helps to study the distribution of climatic parameters in vegetation and the interaction of crop and environment. It also shows the effect of climatic conditions such as temperature, humidity, carbon dioxide concentration, and light conditions on photosynthesis, dry matter accumulation, crop yield, and distribution of fungal diseases in plant populations [68–72].

Szanto [73] determined the optimal row orientation of greenhouse tomato with special emphasis on stomatal conductance, its dynamics, thus assimilation and transpiration, using functional-structural plant modeling. A coupled steady-state photosynthesis and stomatal conductance model was used to evaluate the effects of row orientations on crop performance. Functional-structural plant model of greenhouse tomato was established in GroIMP 1.5, using ray tracing for light environment simulations. Results showed that stomatal dynamics may be a significant reduction factor of assimilation. The vertical distribution of photosynthesis and transpiration did not show discrepancies between the row orientations. The diurnal pattern of assimilation demonstrated that at low solar elevation angles, the direct irradiance should reach the canopy in parallel with the rows, while at higher solar elevation angles, row orientation has a weak effect on the light interception.

Buck-Sorlin et al. [74] used structural plant modeling (FSPM) including a virtual greenhouse environment with the crop, light sources (diffuse and direct sunlight and lamps) and photosynthetically active radiation (PAR) sensors to better understand the processes that help produce the quality and quantity of roses. The crop model is designed as a multiscale FSPM with plant organs (axillary buds, leaves, internodes, flowers) as basic units, and local light interception and photosynthesis within each leaf. The model was able to reproduce PAR measurements at different canopy positions, times of the day, and light conditions. For different typical cultivation scenarios, the simulated incident and the adsorbed PAR, and the net uptake rate in upright and bent shoots showed characteristic spatial and daily dynamics.

Wiechers et al. [75] investigated the effect of the distribution of environmental factors and canopy architecture on growth imbalances between individual fruits of cucumber (*Cucumis sativus*) in a greenhouse. They used the formalism of the L-system to create FSPM, which combined a plant three-dimensional structure model, a photosynthetic biochemical model, and an adsorption model involving fruit growth based on potential growth rate (RP), abortion and dominance. Simulations were performed for a dense row with sparse symmetrical canopies. The results showed that simple partitioning models unsuccessful in simulating the growth of individual fruits. The model had good results in determining the abortion and dominance threshold. There was good agreement in simulating the duration of fruit growth, abortion rate with measurements and reproduction of conditions in which the fruit could be harvested earlier.

Zhang [70] conducted experiments in greenhouses with cut-flower plants (lilies and roses) to determine the response of plants (including response to leaf photosynthetic traits and plant architectural traits) to changes in PAR, R: FR, water level and nitrogen, and study the presence of bent shoots. Then, modeling studies were performed to quantify the photosynthetic response to these conditions at the leaf, plant and crop levels. The results showed that to quantify the effects of environmental factors, plant responses, and biotic processes on crop yield, the combination of the FSP model with detailed leaf photosynthesis models (for both steady-state and dynamic photosynthesis) and phylloclimate models can be used.

Zhang et al. [72] introducing a new method for evaluating micro-light climate and thermal performance in a Liaoshen-type Solar Greenhouse (LSG) incorporated 3D architecture tomato canopy, simulated using an FSP model. The exact surface temperature of each component of the greenhouse and tomato crop was simulated using advanced light modeling techniques. Considering the simulated light absorption as input, the thermal conditions were obtained using particular energy balance equations. Results showed that simulated greenhouse temperatures from cover, ceiling, indoor, wall, canopy and soil had a good agreement with the experimental data.

7. Future trends

Evers et al. [76] are conducting a study on a digital twin greenhouse that aims to develop a simulation model that predicts tomato plant growth in 3D (**Figure 4**). The model simulates crop yield, CO_2 uptake, and use of nutrients, energy, and water, as well as profit and environmental impact. Simulations are based on real-time measurements of tomato plants and their growing conditions. The core of the study is based on the concepts of functional-structural plant (FSP) modeling. The environmental variables driving plant growth and development will be simulated by a greenhouse module based on the Kaspro model. Data from several sensors such as the multi-spectral 3D laser scanner, chlorophyll fluorescence camera, thermal camera, and climate sensors, will be processed to estimate plant traits and climate conditions. The focus is therefore on estimating plant traits from raw sensor data. Based on the model predictions, crop management strategy can be adjusted, and improved plant traits can be identified. The future trend in greenhouse crop production [77] is toward the use of digital technology and robotics [78], artificial intelligence, and collecting greenhouse climate data using IoT sensors [79] in

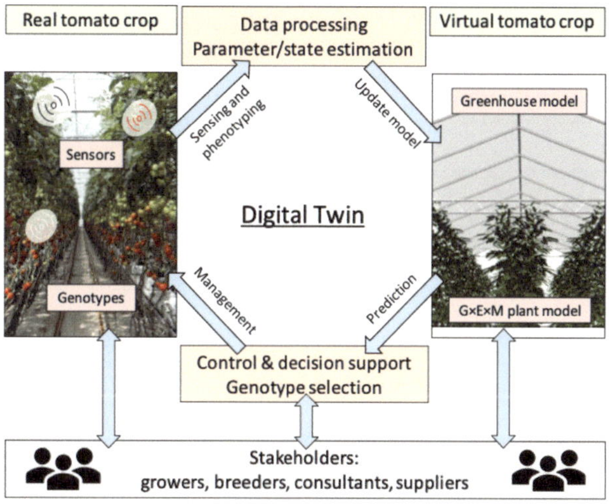

Figure 4.
Concept map of the Virtual Tomato Crops digital twin. (G × E × M = interaction between genotype, environment and management) [76].

combination with FSP growth models and analyzing them in the cloud for preparing a digital version of a real greenhouse.

8. Conclusion

To understand the complexities of the crop responses to its environment and management practices as well as the dynamics of the greenhouse environment in response to external conditions, greenhouse characteristics, and management, the researchers have been developing models of greenhouse environments and crops. The most prominent models of climate dynamics based on energy balance are KASPRO and Vanthoor models. The greenhouse process (KASPRO) model is constructed from modules describing the physics of mass and energy transport in the greenhouse enclosure and the large number of modules that simulate the customary greenhouse climate controllers. The climate controller of KASPRO enables climate management using heating, ventilation, dehumidification, moistening, shading, artificial illumination, and carbon dioxide supply. Vanthoor [47] developed a model to study the effects of outdoor climate and greenhouse design on the indoor greenhouse climate. Extensive studies by computational fluid dynamics have been performed on the effect of ventilation vents arrangement, climate control equipment, greenhouse dimensions, solar radiation, and crop canopy on airflow and indoor climate of the greenhouse. The crop growth model is an essential part of the optimization of cultivation management. Plant dynamics models are often designed for specific conditions and their application in different conditions reduces the accuracy of their results. TOMGRO and Vanthoor are two widely used tomato growth models. With the development of computers, FSP models simulate growth and morphology of individual plants and their interaction with the environment, . from which the complex properties of the plant community emerge. Functional – structural plant models (FSPM) have been used to simulate tomatoes, cucumbers, and roses in the greenhouse. The digital twin greenhouse is currently being studied and developed using IoT sensors, climate models (KASPRO), and FSP models.

Author details

Seyed Moin-E-Ddin Rezvani[1], Redmond R. Shamshiri[2]*, Ibrahim A. Hameed[3], Hamid Zare Abyane[4], Mohsen Godarzi[5], Davood Momeni[6] and Siva K. Balasundram[7]

1 Agricultural Engineering Research Department, Hamedan Agricultural and Natural Resources Research and Education Center, AREEO, Hamedan, Iran

2 Leibniz Institute for Agricultural Engineering and Bioeconomy, Potsdam-Bornim, Germany

3 Department of ICT and Natural Sciences, Faculty of Information Technology and Electrical Engineering, NTNU, Ålesund, Norway

4 Department of Irrigation and Drainage, Faculty of Agriculture, Bu-Ali Sina University, Hamadan, Iran

5 Department of Mechanical Engineering, Faculty of Engineering, Bu-Ali Sina University, Hamadan, Iran

6 Agricultural Engineering Research Department, Isfahan Agricultural and Natural Resources Research and Education Center, AREEO, Isfahan, Iran

7 Department of Agriculture Technology, Faculty of Agriculture, Universiti Putra Malaysia, Serdang, Selangor, Malaysia

*Address all correspondence to: rshamshiri@atb-potsdam.de

References

[1] Jones, J.; Kenig, A.; Vallejos, C. Reduced state–variable tomato growth model. Transactions of the ASAE. 1999; 42: 255 p.

[2] Luo, W.; de Zwart, H.F.; Dail, J.; Wang, X.; Stanghellini, C.; Bu, C. Simulation of greenhouse management in the subtropics, Part I: Model validation and scenario study for the winter season. Biosystems engineering. 2005; 90: 307-318 p.

[3] Piscia, D. Analysis of night-time climate in plastic-covered grenhouses. 2012: p.

[4] Singh, A.K. Crop growth simulation models. IASRI, New Delhi. 1994: 497-509 p.

[5] Lin, D.; Wei, R.; Xu, L. An Integrated Yield Prediction Model for Greenhouse Tomato. Agronomy. 2019; 9: 873 p.

[6] Vos, J.; Marcelis, L.; Evers, J. Functional-structural plant modelling in crop production: adding a dimension. Frontis. 2007: 1-12 p.

[7] López-Cruz, I.; Fitz-Rodríguez, E.; Salazar-Moreno, R.; Rojano-Aguilar, A.; Kacira, M. Development and analysis of dynamical mathematical models of greenhouse climate: A review. European Journal of Horticultural Science. 2018; 83: 269-279 p.

[8] Watt, J. 3D Crop Modelling [thesis]. London: University College London; 2013.

[9] Flores-Velázquez, J.; Rojano, A.; Rojas-Rishor, A.; Bustamante, W.O. Computational Fluid Dynamics Achievements Applied to Optimal Crop Production in a Greenhouse. New Perspectives in Fluid Dynamics. 2015: 77 p.

[10] Hemming, S.; de Zwart, F.; Elings, A.; Petropoulou, A.; Righini, I. Cherry Tomato Production in Intelligent Greenhouses—Sensors and AI for Control of Climate, Irrigation, Crop Yield, and Quality. Sensors. 2020; 20: 6430 p.

[11] Mohammadi, B.; Ranjbar, S.F.; Ajabshirchi, Y. Application of dynamic model to predict some inside environment variables in a semi-solar greenhouse. Information processing in agriculture. 2018; 5: 279-288 p.

[12] Bot, G.P. Greenhouse climate: from physical processes to a dynamic model [thesis]. Bot; 1983.

[13] Van Henten, E. Greenhouse climate management: an optimal control approach; Van Henten; 1994; p.

[14] Taki, M.; Ajabshirchi, Y.; Ranjbar, S. F.; Rohani, A.; Matloobi, M. Modeling and experimental validation of heat transfer and energy consumption in an innovative greenhouse structure. Information Processing in Agriculture. 2016; 3: 157-174 p., doi:https://doi.org/10.1016/j.inpa.2016.06.002.

[15] Boulard, T.; Wang, S. Experimental and numerical studies on the heterogeneity of crop transpiration in a plastic tunnel. Computers and Electronics in agriculture. 2002; 34: 173-190 p.

[16] Bartzanas, T.; Boulard, T.; Kittas, C. Effect of vent arrangement on windward ventilation of a tunnel greenhouse. Biosystems Engineering. 2004; 88: 479-490 p.

[17] Madjoubi, H.; Boulard, T.; Fatnassi, H.; Senhaji, A.; Elbahi, S.; Demrati, H.; Mouqallid, M.; Bouirden, L. Canary greenhouse CFD nocturnal climate simulation. 2016: p.

[18] Majdoubi, H.; Boulard, T.; Fatnassi, H.; Bouirden, L. Airflow and

microclimate patterns in a one-hectare Canary type greenhouse: an experimental and CFD assisted study. Agricultural and Forest Meteorology. 2009; 149: 1050-1062 p.

[19] Fatnassi, H.; Boulard, T.; Poncet, C.; Chave, M. Optimisation of greenhouse insect screening with computational fluid dynamics. Biosystems Engineering. 2006; 93: 301-312 p.

[20] Chen, J.; Xu, F.; Tan, D.; Shen, Z.; Zhang, L.; Ai, Q. A control method for agricultural greenhouses heating based on computational fluid dynamics and energy prediction model. Applied Energy. 2015; 141: 106-118 p.

[21] Jarvis, P. The interpretation of the variations in leaf water potential and stomatal conductance found in canopies in the field. Philosophical Transactions of the Royal Society of London. B, Biological Sciences. 1976; 273: 593-610 p.

[22] Fidaros, D.; Baxevanou, C.; Bartzanas, T.; Kittas, C. Numerical simulation of thermal behavior of a ventilated arc greenhouse during a solar day. Renewable Energy. 2010; 35: 1380-1386 p.

[23] Ali, H.B.; Bournet, P.-E.; Cannavo, P.; Chantoiseau, E. Development of a CFD crop submodel for simulating microclimate and transpiration of ornamental plants grown in a greenhouse under water restriction. Computers and Electronics in Agriculture. 2018; 149: 26-40 p.

[24] Kacira, M.; Sase, S.; Okushima, L. Effects of side vents and span numbers on wind-induced natural ventilation of a gothic multi-span greenhouse. Japan Agricultural Research Quarterly: JARQ. 2004; 38: 227-233 p.

[25] Baeza, E.; Perez-Parra, J.; Lopez, J.; Montero, J. CFD simulation of natural ventilation of a parral greenhouse with a baffle device below the greenhouse vents. In Proceedings of International Symposium on High Technology for Greenhouse System Management: Greensys 2007 801; 885-892 p.

[26] Flores-Velázquez, J.; Montero, J.; Baeza, E.; Lopez, J.; Pérez-Parra, J.; Bonachela, S. Analysis of mechanical ventilation in a three-span greenhouse using computational fluid dynamics (CFD). In Proceedings of International Symposium on High Technology for Greenhouse Systems: GreenSys 2009 893; 653-660 p.

[27] Flores-Velazquez, J.; Montero, J.I.; Baeza, E.J.; Lopez, J.C. Mechanical and natural ventilation systems in a greenhouse designed using computational fluid dynamics. International Journal of Agricultural and Biological Engineering. 2014; 7: 1 p.

[28] Roy, J.; Fatnassi, H.; Boulard, T.; Pouillard, J.-B.; Grisey, A. CFD determination of the climate distribution in a semi closed greenhouse with air cooling. In Proceedings of International Symposium on New Technologies and Management for Greenhouses-GreenSys 2015 1170; 103-110 p.

[29] Flores-Velázquez, J.; Vega-García, M. Regional management of the environment in a zenith greenhouse with computational fluid dynamics (CFD). Ingeniería Agrícola y Biosistemas. 2019; 11: p.

[30] Molina-Aiz, F.; Valera, D.; Pena, A.; Alvarez, A.; Gil, J. Analysis of the effect of rollup vent arrangement and wind speed on Almería-type greenhouse ventilation performance using computational fluid dynamics. In Proceedings of International Symposium on Greenhouse Cooling 719; 173-180 p.

[31] Bournet, P.; Khaoua, S.O.; Boulard, T.; Migeon, C.; Chasseriaux, G. Effect of roof and side opening combinations on

the ventilation of a greenhouse using computer simulation. Transactions of the ASABE. 2007; 50: 201-212 p.

[32] Bartzanas, T.; Katsoulas, N.; Kittas, C.; Boulard, T.; Mermier, M. The effect of vent configuration and insect screens on greenhouse microclimate. International Journal of Ventilation. 2005; 4: 193-202 p.

[33] Majdoubi, H.; Boulard, T.; Hanafi, A.; Fatnassi, H.; Demrati, H.; Bekkaoui, A.; Nya, M.; Bouirden, L. Determination and analysis of air exchange rate in a large greenhouse equipped with insect proof net. Acta horticulturae. 2007; 747: 151 p.

[34] Chen, J.; Cai, Y.; Xu, F.; Hu, H.; Ai, Q. Analysis and optimization of the fan-pad evaporative cooling system for greenhouse based on CFD. Advances in Mechanical Engineering. 2014; 6: 712740 p.

[35] Fidaros, D.; Baxevanou, C.; Bartzanas, T.; Kittas, C. Flow Characteristics and temperature patterns in a fan ventilated greenhouse. In Proceedings of International Workshop on Greenhouse Environmental Control and Crop Production in Semi-Arid Regions 797; 123-129 p.

[36] Baxevanou, C.; Fidaros, D.; Bartzanas, T.; Kittas, C. Numerical simulation of solar radiation, air flow and temperature distribution in a naturally ventilated tunnel greenhouse. Agricultural Engineering International: CIGR Journal. 2010; 12: 48-67 p.

[37] Ortíz-Vázquez, I.C.; Irene, R.-M.L.; Pérez-Robles, J.F.; Soto-Zarazúa, G.; Rico-García, E.; De la Torre-Gea, G.A. Analysis of large commercial greenhouses in warm climates using cfd and bayesian networks. Journal of Global Ecology and Environment. 2016; 5: 91-96 p.

[38] Boulard, T.; Roy, J.-C.; Pouillard, J.-B.; Fatnassi, H.; Grisey, A. Modelling of micrometeorology, canopy transpiration and photosynthesis in a closed greenhouse using computational fluid dynamics. Biosystems Engineering. 2017; 158: 110-133 p.

[39] Rezvani, S.M.-e. Optimization of ventilation and simulation of transpiration in Hamedan greenhouses with computational fluid dynamics and energy balance [thesis]. Iran, Hamedan: Bu-Ali Sina; 2020.

[40] Roy, J.C.; Boulard, T.; KITTAS, C.; Wang, S. Convective and ventilation transfers in greenhouses, part 1: the greenhouse considered as a perfectly stirred tank. Biosystems Engineering. 2002; 83: 1-20 p.

[41] Piscia, D.; Montero, J.I.; Baeza, E.; Bailey, B.J. A CFD greenhouse night-time condensation model. Biosystems Engineering. 2012; 111: 141-154 p.

[42] Van Beveren, P.; Bontsema, J.; Van Straten, G.; Van Henten, E. Minimal heating and cooling in a modern rose greenhouse. Applied energy. 2015; 137: 97-109 p.

[43] Van Beveren, P.; Bontsema, J.; Van Straten, G.; Van Henten, E. Optimal control of greenhouse climate using minimal energy and grower defined bounds. Applied Energy. 2015; 159: 509-519 p.

[44] Kindelan, M. Dynamic Modeling of Greenhouse Environment. Transactions of the ASABE. 1980; 23: 1232-1239 p.

[45] Bot, G. A validated physical model of greenhouse climate. Engineering and Economic Aspects of Energy Saving in Protected Cultivation 245. 1988: 389-396 p.

[46] De Zwart, H. Analyzing energy-saving options in greenhouse cultivation using a simulation model [thesis].

Wageningen: Wageningen University & Research; 1996.

[47] Vanthoor, B.; Stanghellini, C.; Van Henten, E.J.; De Visser, P. A methodology for model-based greenhouse design: Part 1, a greenhouse climate model for a broad range of designs and climates. Biosystems Engineering. 2011; 110: 363-377 p.

[48] Mobtaker, H.G.; Ajabshirchi, Y.; Ranjbar, S.F.; Matloobi, M. Solar energy conservation in greenhouse: Thermal analysis and experimental validation. Renewable Energy. 2016; 96: 509-519 p.

[49] Yildiz, I.; Stombaugh, D. Dynamic modeling of microclimate and environmental control strategies in a greenhouse coupled with a heat pump system. In Proceedings of III International Symposium on Models for Plant Growth, Environmental Control and Farm Management in Protected Cultivation 718; 331-340 p.

[50] Salazar-Moreno, R.; Irineo, L.; Sánchez Cruz, A.C. Dynamic energy balance model in a greenhouse with tomato cultivation: simulation, calibration and evaluation. Revista Chapingo Serie Horticultura. 2019; 25: p.

[51] Joudi, K.A.; Farhan, A.A. A dynamic model and an experimental study for the internal air and soil temperatures in an innovative greenhouse. Energy Conversion and Management. 2015; 91: 76-82 p.

[52] Challa, H. Crop models for greenhouse production systems. In Proceedings of IV International Symposium on Models for Plant Growth and Control in Greenhouses: Modeling for the 21st Century-Agronomic and 593; 47-53 p.

[53] Jamison, H. Dynamic Modeling Of Tree Growth And Energy Use In A Nursery Greenhouse Using Matlab And Simulink; 2006.

[54] Shamshiri, R. Measuring optimality degrees of microclimate parameters in protected cultivation of tomato under tropical climate condition. Measurement. 2017; 106: 236-244 p.

[55] Shamshiri, R.R.; Mahadi, M.R.; Thorp, K.R.; Ismail, W.I.W.; Ahmad, D.; Man, H.C. Adaptive management framework for evaluating and adjusting microclimate parameters in tropical greenhouse crop production systems. Plant Engin. 2017: 167-191 p.

[56] Shamshiri, R.R.; Jones, J.W.; Thorp, K.R.; Ahmad, D.; Che Man, H.; Taheri, S. Review of optimum temperature, humidity, and vapour pressure deficit for microclimate evaluation and control in greenhouse cultivation of tomato: a review. International agrophysics. 2018; 32: 287-302 p.

[57] Heuvelink, E. Evaluation of a dynamic simulation model for tomato crop growth and development. Annals of Botany. 1999; 83: 413-422 p.

[58] Abreu, P.; Meneses, J.; Gary, C. Tompousse, a model of yield prediction for tomato crops: calibration study for unheated plastic greenhouses. In Proceedings of XXV International Horticultural Congress, Part 9: Computers and Automation, Electronic Information in Horticulture 519; 141-150 p.

[59] Jones, J.W.; Dayan, E.; Allen, L.; Van Keulen, H.; Challa, H. A dynamic tomato growth and yield model (TOMGRO). Transactions of the ASAE. 1991; 34: 663-0672 p.

[60] Dimokas, G.; Tchamitchian, M.; Kittas, C. Calibration and validation of a biological model to simulate the development and production of tomatoes in Mediterranean greenhouses during winter period. biosystems engineering. 2009; 103: 217-227 p.

[61] Gallardo, M.; Thompson, R.; Rodríguez, J.; Rodriguez, F.; Fernández,

M.; Sánchez, J.; Magán, J. Simulation of transpiration, drainage, N uptake, nitrate leaching, and N uptake concentration in tomato grown in open substrate. Agricultural Water Management. 2009; 96: 1773-1784 p.

[62] Vanthoor, B.; Van Henten, E.; Stanghellini, C.; De Visser, P. A methodology for model-based greenhouse design: Part 3, sensitivity analysis of a combined greenhouse climate-crop yield model. Biosystems engineering. 2011; 110: 396-412 p.

[63] Evers, J.B.; Letort, V.; Renton, M.; Kang, M. Computational botany: advancing plant science through functional–structural plant modelling. Oxford University Press US: 2018.

[64] Evers, J.B. Simulating crop growth and development using functional-structural plant modeling. In Canopy photosynthesis: from basics to applications, Springer; 2016. 219-236 p. ^pp.

[65] Louarn, G.; Song, Y. Two decades of functional–structural plant modelling: now addressing fundamental questions in systems biology and predictive ecology. Annals of Botany. 2020; 126: 501-509 p.

[66] Vanthoor, B.; De Visser, P.; Stanghellini, C.; Van Henten, E. A methodology for model-based greenhouse design: Part 2, description and validation of a tomato yield model. Biosystems engineering. 2011; 110: 378-395 p.

[67] Vanthoor, B.H.; Gazquez, J.C.; Magan, J.J.; Ruijs, M.N.; Baeza, E.; Stanghellini, C.; van Henten, E.J.; de Visser, P.H. A methodology for model-based greenhouse design: Part 4, economic evaluation of different greenhouse designs: A Spanish case. biosystems engineering. 2012; 111: 336-349 p.

[68] Guo, Y.; Fourcaud, T.; Jaeger, M.; Zhang, X.; Li, B. Plant growth and architectural modelling and its applications. Annals of botany. 2011; 107: 723-727 p.

[69] Vos, J.; Evers, J.B.; Buck-Sorlin, G. H.; Andrieu, B.; Chelle, M.; De Visser, P. H. Functional–structural plant modelling: a new versatile tool in crop science. Journal of experimental Botany. 2010; 61: 2101-2115 p.

[70] Zhang, N. From leaf to crop: quantifying photosynthesis responses of two flower crops [thesis]. Wageningen University; 2019.

[71] Zhang, N.; Van Westreenen, A.; Evers, J.B.; Anten, N.P.; Marcelis, L.F. Quantifying the contribution of bent shoots to plant photosynthesis and biomass production of flower shoots in rose (Rosa hybrida) using a functional–structural plant model. Annals of botany. 2020; 126: 587-599 p.

[72] Zhang, Y.; Henke, M.; Li, Y.; Yue, X.; Xu, D.; Liu, X.; Li, T. High resolution 3D simulation of light climate and thermal performance of a solar greenhouse model under tomato canopy structure. Renewable Energy. 2020; 160: 730-745 p.

[73] Szanto, C. In search of the optimal greenhouse tomato row orientation using functional-structural plant modeling, with focus on stomatal conductance [thesis]. Netherlands: Wageningen Agricultural University; 2016.

[74] Buck-Sorlin, G.; de Visser, P.H.; Henke, M.; Sarlikioti, V.; van der Heijden, G.W.; Marcelis, L.F.; Vos, J. Towards a functional–structural plant model of cut-rose: simulation of light environment, light absorption, photosynthesis and interference with the plant structure. Annals of Botany. 2011; 108: 1121-1134 p.

[75] Wiechers, D.; Kahlen, K.; Stützel, H. Dry matter partitioning models for the

simulation of individual fruit growth in greenhouse cucumber canopies. Annals of Botany. 2011; 108: 1075-1084 p.

[76] Evers, J. The Digital Twin project Virtual Tomato Crops (VTC). Availabe online: https://www.wur.nl/en/show/The-Digital-Twin-project-Virtual-Tomato-Crops.htm (accessed on 15/01/2015).

[77] R Shamshiri, R.; Kalantari, F.; Ting, K.; Thorp, K.R.; Hameed, I.A.; Weltzien, C.; Ahmad, D.; Shad, Z.M. Advances in greenhouse automation and controlled environment agriculture: A transition to plant factories and urban agriculture. 2018: p.

[78] R Shamshiri, R.; Weltzien, C.; Hameed, I.A.; J Yule, I.; E Grift, T.; Balasundram, S.K.; Pitonakova, L.; Ahmad, D.; Chowdhary, G. Research and development in agricultural robotics: A perspective of digital farming. 2018: p.

[79] Shamshiri, R.R.; Bojic, I.; van Henten, E.; Balasundram, S.K.; Dworak, V.; Sultan, M.; Weltzien, C. Model-based evaluation of greenhouse microclimate using IoT-Sensor data fusion for energy efficient crop production. Journal of Cleaner Production. 2020; 263: 121303 p.

Chapter 7

Combating Greenhouse Effects through Biomass Gasification: A Focus on Kinetic Modeling of Combustion and Gasification Zones

Sunday J. Ojolo and Musbau G. Sobamowo

Abstract

The prevalent challenges of global warming, food security, food production, crop production systems, environment control called for consideration and better utilization of green energy system such as biomass. The advanced thermo-chemical conversion of the renewable energy source which is aimed at production of optimal yield of energy has not been well understood. In order to have better physical insights into the detailed structure of the biomass burning process inside a solid bed, the kinetics of the biomass combustion and gasification must be properly analyzed. Consequently, improved kinetic models of the combustion and gasification zones in the thermochemical conversion system are very required. Therefore, the present study focuses on the development of improved kinetic modeling of the combustion and gasification zones in the biomass gasification system. The performance of the biomass gasifier system is evaluated through the equivalence ratio, the syngas composition, cold gas efficiency and lower heating value. Also, the effects of the equivalent ratio on gas compositions, the gasifier performance and the low heating value of the biomass are analyzed. From the analysis, it is established that the concentration of CO, H_2 and CH_4 in the gasifier decrease as the equivalence ratio increases. However, CO_2 concentration increases with an increase in the equivalence ratio. The cold efficiency and LHV decreases as the equivalence ratio increases while the gas yield increases with an increase in the equivalence ratio. The quantity of gas produced increases as the amount of oxygen consumed increases. Also, the ratio of CO/CO_2 decreases as the temperature of the reduction zone increases. Such analysis as presented in this work, is very useful as a time-saving and cost-effective tool for designing and optimizing the biomass gasifier. Therefore, it is evident that this work will play a significant role in the system design including analysis of the distribution of products and ash deposit in the downdraft gasifiers.

Keywords: Kinetic models, Combustion Zone, Gasification Zone, Downdraft Biomass Gasifier

1. Introduction

In the past few decades, the increasing concerns of global warming and fuel prices have aroused the development of new technologies in alternative energy. However, meeting the future demands for electricity, heat, cooling, fuels, and materials with highly limited and fluctuating resources, requires careful planning and allocation of the available resources with highly flexible systems. One of the few renewable resources that is capable of supplying the needs is biomass energy. Biomass as a source of renewable energy and as an organic material from plants and animals can be biochemically and thermochemically converted to produce heat, electricity and fuels. Among all the biomass conversion processes, gasification is one of the most promising [1]. Biomass gasification allows an environmentally friendly energy production. In such a thermochemical conversion process of solid fuel, the most important properties relating to the thermal and biological conversions are moisture content, ash content, volatile matter, and energy density. Therefore, an assessment of the use of biomass as a fuel requires a basic understanding of their composition, characteristics, and performance. The performance of the renewable energy sources in the combustion and gasification systems is ultimate determined by it specific properties [2]. Since biomass materials exhibit a wide range of moisture contents which affect their low and high heating values as a fuel source, it is very important to establish the moisture content of the biomass materials. High or excessive moisture content could defeat the main purpose of the biomass gasification process [3]. Also, the amount of the inorganic component (ash content) in biomass is very important to be determined especially for temperature gasification as melted ash may cause problems in the reactor [2]. The effects of moisture and ash contents on the low heating value (LHV) of some types of biomass are shown in **Table 1**.

The thermal conversion process which involves incomplete combustion of biomass due to insufficient amounts of oxygen from the available supply of air, produces synthetic gases (syngas). Although, the actual biomass syngas composition depends on the gasification process, the feedstock composition and the gasifying agent, a typical syngas by weight from gasification of wood contains approximately 15–21% hydrogen (H_2), 10–20% carbon monoxide (CO), 11–13% carbon dioxide, and 1–5% of methane which are combustible gases. The nitrogen gas (N_2) involved in the gasification process is not combustible but it dilutes the syngas as it enters and burns in an engine. Compared to biomass combustion, biomass gasification has a lower environmental impact due to less greenhouse gas emission. Therefore, biomass gasification has been beneficial in decreasing greenhouse gases emissions. The reduction of fossil fuels dependence for energy supply, the decrease of land use and soil contamination for waste disposal, the higher efficiency and reliability of a

Biomass type (%)	Moisture content (%)	Ash content (dry) (kJ/kg)	Lower heat value
Wood	10–60	0.25–1.70	8,400-17,000
Bagasse	40–60	1.70–3.80	7,700-8,000
Stalk	10–20	0.10	16,000
Rice husks	9	19.00	14.000
Gin trash	9	12.00	14,000

Source: Quaak et al. *[2].*

Table 1.
Effects of moisture and ash contents on LHV of some types of biomass.

grid-connected power net, and, on a larger scale, contribution to air pollution control and global warming reduction are the reasons for the increasing utilization of the biomass gasification technologies.

Indisputably, the optimal yield of synthetic gases from gasifiers has been the main focus of the thermochemical conversion technologies. Based on the method of air introduction, solid fuel usage in the gasification zone and the direction of the syngas leaving the gasifier, there are four types of gasifiers, namely, updraft or countercurrent gasifiers; downdraft or co-current gasifiers; crossdraft gasifiers; and fluidized-bed gasifiers. These four types of gasifiers can be broadly classified as fixed and fluidized bed gasifiers. Yang et al. [4] reported that fixed bed gasification is the most common technology for the energy use of biomass and solid municipal wastes. They are relatively easy to design and operate but have limited capacity. Therefore, fixed bed gasifiers are preferred for small to medium scale applications with thermal requirements up to 1 MW [5, 6]. Fixed bed gasifiers include updraft and downdraft gasifiers The updraft gasifier comes with simple design,, high charcoal burn-out and internal heat exchange. Such reactor has low gas exit temperatures and. However, in such reactor, there is possibility of "channeling" in the equipment, which can lead to oxygen break-through and explosion. The requirement of installing automatic moving grates coupled with the problems associated with the disposal of the tar-containing condensates that result from the gas cleaning operations are also some of the major setbacks in the wide applications of the type of gasifier. **Table 2** shows the Typical Characteristics of Fixed-Bed and Fluidized-Bed Gasifiers.

Downdraft gasification is a comparatively cheap method of gasification that can yield producer gas with very low tar content. The downdraft gasifier has a simple and stable design, making it effective for small and modular applications if it is well designed. However, downdraft gasifiers cannot be used in some unprocessed fuels. Such gasifier produces higher ash content fuels (slagging) than updraft gasifiers. Also, fluffy, low-density materials give rise to flow problems and excessive pressure drop, and the solid fuel must be pelletized or briquetted before use. As compared to updraft gasifier, downdraft gasifier has lower efficiency due to the lack of internal heat exchange and the lower heating value of the gas. Also, it can only be used in a power range of less than 1 MW due to the necessity of maintaining uniform high temperatures over a given cross-sectional area. The operation of the fixed bed

Characteristics	Fixed-bed downdraft	Fluidized-bed gasifier
Fuel size(mm)	10–200	0–20
Ash content (%wt)	< 6	< 25
Moisture content	> 10, < 25	> 30
Operating temperature (°C)	800–1400	750–950
Control	Simple	Average
Turndown ratio	4	3
Capacity (MW)	< 2.5	< 1–50
Hot gas efficiency (full load %)	85–90	—
Cold gas efficiency (full load %)	65–75	—
Tar content (g/Nm^3)	< 0.5	< 1.5
Low Heating Value (kJ/Nm^3)	4.5–5.0	1.0

Table 2.
Typical characteristics of fixed-bed and fluidized-bed gasifiers.

gasifiers is influenced by the morphological, physical and chemical properties of the fuel. In such reactors, there are technical problems such as lack of bunker flow, slagging and extreme pressure drop over the gasifier.

In order to combat these problems, a fluidized bed gasifier was developed. The fluidized-bed gasifiers are able to handle a wide range of biomass with high throughput. However, the fluidized bed design presents several flaws such as particle fracture due to collision with the walls of the vessel, propeller blades, and adjacent particles that render the packing ineffective and useless. Maintenance costs associated with the moving parts increase the overhead needed to repair the damage. These flaws make the immobilized packed design (fixed bed design) a more attractive and practical alternative since it eliminates the moving parts and their inherent problems, and also, allows for packing or media to be simply regenerated once it becomes saturated with contaminants. Instability of the reaction bed, feeding problems and fly-ash sintering in the gas channels can occur in fluidized bed. Additionally, the fluidized bed gasifier produces high tar content of the product gas (up to 500 mg/m^3 gas) with incomplete carbon burn-out, and poor response to load changes. A solution to the problem of tar entrainment in the gas stream is provided by the utilizations of downdraft gasifiers. In fact, the downdraft gasifiers have the possibility of producing a tar-free gas suitable for engine applications. Because of the lower level of organic components in the condensate, downdraft gasifiers suffer less from environmental objections than updraft and fluidized gasifiers.

The understanding of the interactions between chemical and physical mechanisms occurring in the gasifier is of fundamental importance for the optimal design of the reactor. This therefore, provokes the simulations of the thermochemical processes in the gasifiers. The increased computer efficacy and advanced numerical techniques as possessed in various numerical simulation techniques such as CFD tools have offered an effective means of simulating the physical and chemical processes in the biomass thermo-chemical reactors (such as fluidized beds, fixed beds, combustion furnaces, firing boilers, rotating cones and rotary kilns) under various operating conditions in different virtual environments. The CFD simulates the fluid flow behavior, heat and mass transfer, chemical reactions, phase changes and mechanical movement. CFD model results are capable of predicting both qualitative and quantitative information. The results of accurate simulations with the aid of CFD tools can help to optimize the system design and operations and understand the dynamic processes inside the reactors. Also, the use of CFD software for the flow visualization in a fixed bed gasifier has resulted in saving stresses and time engages in using other modeling and simulation methods. Additionally, the predictions made by the use of CFD software can help in the design of the fixed bed gasifier. Consequently, Fletcher et al. [7] simulated the flow and reaction in an entrained flow biomass gasifier using the CFX package. With the aid of CFD model, Gerun et al. [8] presented a two-dimensional heat and mass transfer for the oxidation zone in a two-stage downdraft gasifier. Meanwhile, Zhou et al. [9] adopted a CFD model to simulate the optimal conditions for the production of hydrogen-rich gas in a fluidized bed biomass gasifier. A further study was presented by Papadikis and Gu [10]. The authors presented CFD modeling of the fast pyrolysis of biomass in fluidized bed reactors. In the same year, Wang and Yan [11] carried out an overview of different CFD studies on thermo-chemical biomass conversions such as gasification and combustion processes in fixed beds, furnaces, and fluidized beds.

In some specific studies, Yimlaz, et al. [12] and Cornejo and Farias [13] adopted the multiphase model in FLUENT while Paes [14], Watanabe and Otaka [15], Huang and Ramaswamy [16] and with the aid of mathematical user-defined functions, Cuoci, et al. [17] presented the mathematical model of the thermochemical processes. Focusing on using oil-palm fronds, Atnaw and Sulaiman [18] submitted a modeling and simulation study of downdraft gasifier. With the aid of computational fluid dynamics techniques, Hamzehei et al. [19] adopted multi-fluid Eulerian modeling while incorporating the kinetic theory for solid particles to simulate the unsteady-state behavior of two-dimensional non-reactive gas–solid fluidized bed reactor. The CFD tool was utilized by Tingwen et al. [20] to present detailed high-resolution simulations of coal injection in a gasifier applying. In the following year, the hydrodynamic behaviors of an internally interconnected fluidized beds (IIFB) which is a novel, self-heating biomass fast pyrolysis reactor was studied by Zhang et al. [21] using the computational tool. The hydrodynamic behavior in a circulating fluidized bed (CFB) riser was studied by Peng et al. [22] with the help of the CFD model. Chen et al. [23] analyzed a three-dimensional simulation of gas–solid flow in biomass circulating fluidized bed gasifier's riser. Considering the conditions of highly preheated air and steam, Wu et al. [24] showed the a two-dimensional CFD simulation of biomass gasification in a downdraft fixed-bed gasifier. Luo [25] and Liu et al. [26] submitted a three-dimensional CFD model to simulate the full-loop of a dual fluidized-bed biomass gasification system consisting of a gasifier, a combustor, a cyclone separator, and a loop-seal. In the recent year, Lu et al. [27] used CFD model to analyze an updraft gasifier that uses carbonized woody briquette as fuel while in the same year, Kumar et al. [28] investigated the thermochemical conversion of biomass in a downdraft gasifier with a volatile break-up approach. Yang et al. [4] carried out a Eulerian–Lagrangian simulation of air–steam biomass gasification in a three-dimensional bubbling fluidized gasifier.

The predictions of the temperature and pressure distributions and histories of a biomass particle in a downdraft gasifier are major determinants of a detailed characterization study of thermochemical conversions of biomass. Such predictions are heavily dependent on the kinetic modeling of the reaction process in the biomass gasification system [29, 30]. Consequently, improved kinetic models in the combustion and gasification zones in the thermochemical system are very needed. Therefore, the present study focuses on the developing improved kinetic models of the combustion and gasification zones in the biomass gasification system [31, 32]. The performance of the biomass gasifier system is evaluated through the equivalence ratio, the syngas composition, cold gas efficiency and lower heating value (LHV). Also, the effects of the equivalent ratio on gas compositions, the gasifier performance and the low heating value of the biomass are analyzed and presented.

2. The description of downdraft biomass gasifier

Downdraft gasifier has four distinct zones which are drying, pyrolysis, oxidation/combustion, and reduction/gasification zones from top to bottom of the gasifier, respectively. In this type of gasifier, air or oxygen is usually admitted or drawn to the fuel bed in the drying zone through intake nozzles from the throat attached to the combustion zone of the gasifier as shown in **Figure 1a**.

In the **drying zone**, biomass fuel descends into the gasifier and its moisture is removed by evaporation using heat generated in the zones below. The reaction model is shown in Eq. (1) and **Figure 2**.

$$\begin{aligned} &WetBiomass\{CH_\alpha O_\beta(w-y+z)H_2O + m(O_2+3.76N_2)\} \\ &\quad \rightarrow DryBiomass\{CH_\alpha O_\beta(w-y)H_2O + m(O_2+3.76N_2)\} + zH_2O \end{aligned} \tag{1}$$

After the evaporation of free surface water from the biomass as shown in Eq. (1), the dried biomass fuel descends to the pyrolysis **zone** where irreversible

Figure 1.
(a) Downdraft biomass gasification plant. (b) Downdraft biomass gasifier.

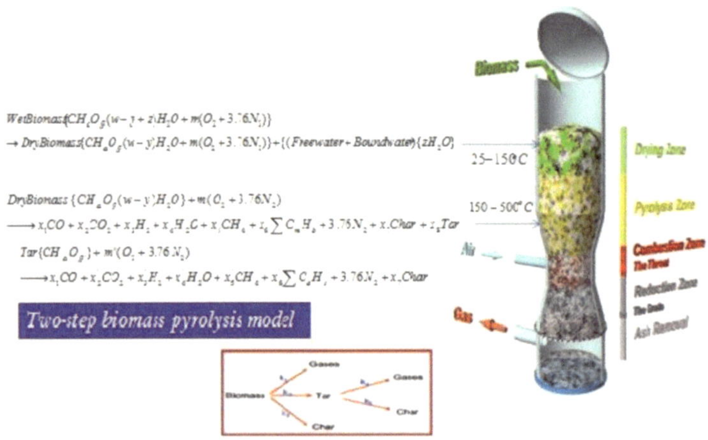

Figure 2.
Reactions at the drying and pyrolysis zones in the downdraft biomass gasifier.

thermal degradation takes place (Eq. (2)). The drying process is achieved by using the released heat energy released from the partial oxidation of the pyrolysis products. When the producer gas flows downwards through the reactor, it enables the pyrolysis gases to pass through hot bed of char which is around 1100°C. Thus, it will crack most of the tars into light chain hydrocarbon and water vapors as shown in Eq. (3) and **Figure 2**.

Primary Pyrolysis

$$\{Biomass \rightarrow (1-\gamma-\Psi)Gases + \Psi Tar + \gamma Char\}$$

$$DryBiomass\{CH_\alpha O_\beta (w-y)H_2O + z(O_2 + 3.76N_2)\}$$
$$\rightarrow x_1CO + x_2CO_2 + x_3H_2 + x_4H_2O + x_5CH_4 + x_6\sum C_\lambda H_\vartheta + 3.76N_2 + x_7Char + x_8Tar \quad (2)$$

Secondary Pyrolysis

$$\{\Psi Tar \rightarrow \lambda Gases + (\Psi-\lambda)Char\}$$

$$Tar\{C_{\alpha'}H_{\beta'}O_{\gamma'}(w' - y')H_2O + z'(O_2 + 3.76N_2)\}$$
$$\rightarrow x_1'CO + x_2'CO_2 + x_3'H_2 + x_4'H_2O + x_5'CH_4 + x_6' \sum C_{\lambda'}H_{\vartheta'} + 3.76N_2 + x_7'Char \tag{3}$$

The Tar is given as $C_{6.407}H_{11.454}O_{3.482}$ while the Char is $CH_{0.2526}O_{0.0237}$.

In the **oxidation/combustion zone**, the volatile products and the char produced during pyrolysis are partially oxidized (Eqs. (4)–(8)) in the exothermic reactions resulting in a rapid rise of temperature up to 1200°C in the throat region. The heat generated is used to drive the gasification reactions.

$$H_2 + \frac{1}{2}O_2 \rightarrow H_2O \tag{4}$$

$$CO + \frac{1}{2}O_2 \rightarrow CO_2 \tag{5}$$

$$CH_4 + 2O_2 \rightarrow CO_2 + 2H_2O \tag{6}$$

$$Char\{CH_aO_b\} + yO_2 \rightarrow \left(2 - 2y - b + \frac{a}{2}\right)CO + \left(2y + b - \frac{a}{2} - 1\right)CO_2 + \frac{a}{2}H_2O \tag{7}$$

$$C_nH_{2n+2} + (2n+1)O_2 \rightarrow nCO + (n+1)H_2O \tag{8}$$

It should be stated that the combustion zone determines the temperatures in the gasifier and the reactions in the other zones and is therefore pivotal in the gasification process. **Figure 3** depicts the combustion reactions in the downdraft gasifier.

The last zone in the downdraft gasifier is the **reduction zone** often refers to as the gasification zone as shown in **Figure 4**. In this zone, the char produced during pyrolysis is converted to the producer gas by reacting with the hot gases from the upper zones. The gases are reduced to form a greater proportion of H_2, CO and CH_4. While leaving the gasifier at the temperature between 200°C and 300°C, the produced gases carry over some dust, pyrolytic products (tar), and water vapor and these are removed by scrubber and electrostatic precipitator.

$$Char\{CH_{a'}O_{b'}\} + (1-b)H_2O \rightarrow CO + \left(1 + \frac{a'}{2} - b'\right)H_2 \tag{9}$$

$$Char\{CH_{a'}O_{b'}\} + CO_2 \rightarrow 2CO + b'H_2O + \left(\frac{a'}{2} - b'\right)H_2 \tag{10}$$

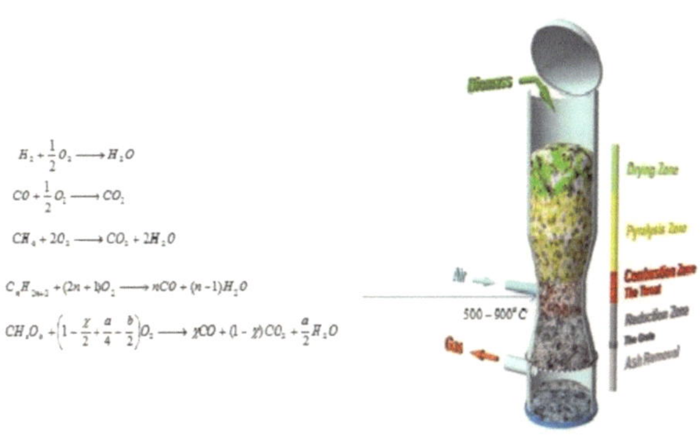

Figure 3.
Reactions at the combustion zone in the downdraft biomass gasifier [33].

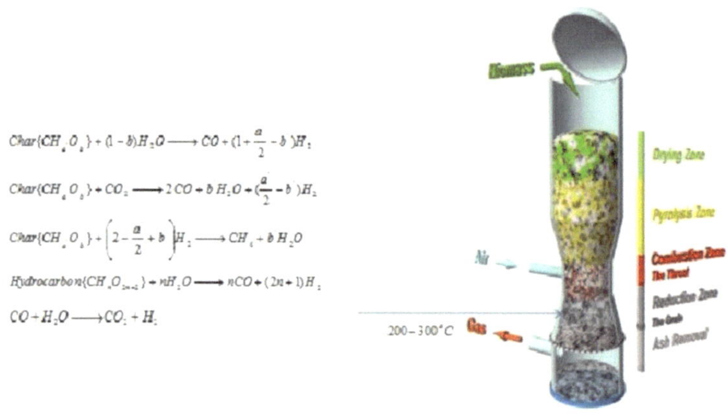

Figure 4.
Reactions at the reduction zone in the downdraft gasifier.

$$Char\{CH_{a'}O_{b'}\} + \left(2 - \frac{a'}{2} + b'\right)H_2 \rightarrow CH_4 + \left(b'H_2O\right) \quad (11)$$

$$Hydrocarbon\{CH_nO_{2n+2}\} + nH_2O \rightarrow nCO + (2n+1)H_2 \quad (12)$$

$$CO + H_2O \rightarrow CO_2 + H_2 \quad (13)$$

where a = 0.2526 and b = 0.0237.

3. Kinetic reactions models for the combustion and the gasification processes

The combustion of CO_2, CO, H_2, CH4 and Hydrocarbon as well as the char combustion and gasification chemical-reactions are determined through improved chemical models. The models involve both homogeneous and heterogeneous reactions. The heterogeneous reaction-rates were determined by combining an Arrhenius kinetic-rate and a diffusion rate using the kinetics/diffusion surface reaction model. Homogeneous reaction-rates were described by a turbulent mixing rate using the eddy dissipation model.

The kinetic reaction rates in the gas phase and of the char are given as.

3.1 Hydrogen combustion

$$r_{H_2-O_2} = \min\left(\left\{2\varepsilon_{H_2} A_{H_2}\ T_e^{-1.5}.\exp\left(\frac{-E_{H_2}}{RT_e}\right)C_{H_2}^{1.5}C_{O_2}\right\},\ C_{mix}\rho_{H_2}\left[\frac{150D_{H_2}(1-\varepsilon_{H_2})^{2/3}}{l_p^2\varepsilon_{H_2}} + \frac{1.75u_{H_2}(1-\varepsilon_{H_2})^{1/3}}{l_p\varepsilon_{H_2}}\min\left[\frac{C_{fuel}}{S_{fuel}}, \frac{C_{O_2}}{S_{O_2}}\right]\right]\right) \quad (14)$$

3.2 Carbon monoxide combustion

$$r_{CO-O_2} = \min\left(\left\{\varepsilon_{CO}\ A_{CO}\exp\left(\frac{-E_{H_2}}{RT_e}\right)C_{CO}C_{O2}^{0.25}C_{H_2O}^{0.5}\right\},\ C_{mix}\rho_{CO}\left[\frac{150D_{CO}(1-\varepsilon_{CO})^{2/3}}{l_p^2\varepsilon_{CO}} + \frac{1.75u_{CO}(1-\varepsilon_{CO})^{1/3}}{l_p\varepsilon_{CO}}\min\left[\frac{C_{fuel}}{S_{fuel}}, \frac{C_{O_2}}{S_{O_2}}\right]\right]\right) \quad (15)$$

3.3 Methane combustion

$$r_{CH_4-O_2} = \min\left(\left\{\varepsilon_{CH_4} A_{CH_4} T_e^{-1}.\ \exp\left(\frac{-E_{CH_4}}{RT_e}\right) C_{CH_4} C_{O_2}\right\},\ C_{mix}\rho_{CH_4}\left[\frac{150 D_{CH_4}(1-\varepsilon_{CH_4})^{2/3}}{l_p^2 \varepsilon_{CH_4}} + \frac{1.75 u_{CH_4}(1-\varepsilon_{CH_4})^{1/3}}{l_p \varepsilon_{CH_4}} \min\left[\frac{C_{fuel}}{S_{fuel}}, \frac{C_{O_2}}{S_{O_2}}\right]\right.\right) \quad (16)$$

3.4 Char combustion

$$r_{char-O_2} = \frac{(A/V)_{char}\varepsilon_{O_2} C_{O_2}\left[\frac{M_{char}}{\left(1-\frac{\chi}{2}+\frac{a}{2}-\frac{b}{2}\right)M_{O_2}}\right]}{\frac{1}{k_d}+\frac{1}{k_k}} \quad (17)$$

where

$$k_k = A_{char}\exp\left(\frac{-E_{char}}{RT_s}\right) \quad (18)$$

3.5 Hydrocarbon combustion

$$r_{C_nH_{2n+2}-O_2} = \min\left(\begin{array}{l}\left\{\varepsilon_{C_nH_{2n+2}} A_{C_nH_{2n+2}} T_e P^{0.3}_{C_nH_{2n+2}} \exp\left(\frac{-E_{H_2}}{RT_e}\right) C^{0.5}_{C_nH_{2n+2}} C_{O_2}\right\}, \\ C_{mix}\rho_{C_nH_{2n+2}}\left[\frac{150 D_{C_nH_{2n+2}}\left(1-\varepsilon_{C_nH_{n+12}}\right)^{2/3}}{l_p^2\varepsilon_{C_nH_{2n+2}}} + \frac{1.75 u_{C_nH_{2+2}}\left(1-\varepsilon_{C_nH_{2n+2}}\right)^{1/3}}{l_p\varepsilon_{C_2H_{2n+2}}} \min\left[\frac{C_{fuel}}{S_{fuel}}, \frac{C_{O_2}}{S_{O_2}}\right]\right.\end{array}\right) \quad (19)$$

where

$$\begin{array}{ll} T_e = \Omega T_g + (1-\Omega)T_s & T_g \le T_s \\ T_e = T_g & T_g \le T_s \end{array} \quad (19a)$$

$$\Omega \textit{ is a weighing factor}$$

3.6 Char reaction with water vapor

$$R_{char-H_2O} = \frac{(A/V)_{char}\varepsilon_{H_2O} C_{H_2O}\left[\frac{M_{Char}}{(1-b)M_{H_2O}}\right]}{\frac{1}{k_d} + \frac{1}{k_r}} \quad (20)$$

where

$$k_k = A_{H_2O} T_{char}\exp\left(\frac{-E_{char}}{RT_{char}}\right) \quad (21)$$

3.7 Char reaction with carbon dioxide

$$R_{char-CO_2} = \frac{(A/V)_{char} C_{CO_2}\left[\frac{M_{Char}}{M_{CO_2}}\right]}{\frac{1}{k_d}+\frac{1}{k_r}} \quad (22)$$

where

$$k_k = A_{CO_2} T_{char}\exp\left(\frac{-E_{char}}{RT_{char}}\right) \quad (23)$$

3.8 Char reaction with hydrogen gas

$$R_{char-H_2} = \frac{(A/V)_{char} C_{H_2} \left[\frac{M_{Char}}{\left(2+b-\frac{a}{2}\right)M_{H_2}}\right]}{\frac{1}{k_d} + \frac{1}{k_r}} \tag{24}$$

where

$$k_k = A_{H_2} \exp\left(\frac{-E_{char}}{RT_{char}}\right) \tag{25}$$

3.9 Water-gas shift

$$R_{CO-H_2O} = \varepsilon_{CO} k_{WGS} \left(C_{CO} C_{H_2O} - \frac{C_{CO_2} C_{H_2}}{k_E} \right) \tag{26}$$

where

$$k_{WGS} = A_{WGS} \exp\left(\frac{-E_{WGS}}{RT_{CO}}\right), k_E = A_E \exp\left(\frac{-E_E}{RT_{CO}}\right) \tag{27}$$

3.10 Steam reforming of hydrocarbon

$$R_{CH_nO_{2n+2}} = A_{sr} T^{b+1} \exp\left(\frac{-E_{sr}}{RT}\right) C_{CH_nO_{2n+2}} \; C_{H_2O} \tag{28}$$

where

$$k_d = \begin{pmatrix} \dfrac{D_g\left(0.644\,\mathrm{Re}^{0.5} Sc^{0.433}\right)}{d_p}, & \textit{Rectangular} - \text{shaped particles} \\ \dfrac{D_g\left(2 + 1.1\,\mathrm{Re}^{0.6} Sc^{1/3}\right)}{d_p} & \textit{Cylindrical} - \text{shaped particles} \\ \dfrac{D_g\left(1.05 + 0.6\,\mathrm{Re}^{0.6} Sc^{1/3}\right)}{d_p} & \textit{Spherical} - \text{shaped particles} \end{pmatrix} \tag{29}$$

Where at a low Reynold number

$$k_d = \frac{D_{H_2}\{W_0(\mathrm{Re})Sc^{0.33} + W_1(\mathrm{Re})\}}{d_p} \tag{30}$$

$$W_o = \begin{bmatrix} 0.4627\exp\left(1.01633(\log_{10}(\mathrm{Re})) + 0.05121(\log_{10}(\mathrm{Re})^2\right. & 10^{-2} \le \mathrm{Re} \le 3.14 \\ 0.597\left(\dfrac{\mathrm{Re}}{In\left(\frac{1}{\mathrm{Re}}\right)^{\frac{1}{3}}}\right) & \mathrm{Re} < 10^{-2} \end{bmatrix} \tag{31}$$

$$W_1 = \begin{bmatrix} \begin{bmatrix} 0.10666\exp\left(0.41285(\log_{10}(\mathrm{Re})) + 0.43847(\log_{10}(\mathrm{Re}))^2 + 0.1915(\log_{10}(\mathrm{Re}))^3\right. \\ +0.01802(\log_{10}(\mathrm{Re}))^4 - 0.005225(\log_{10}(\mathrm{Re}))^5 \end{bmatrix} & 10^{-2} \le \mathrm{Re} \le 3.14 \\ 0.0917 & Re < 10^{-2} \end{bmatrix} \tag{32}$$

4. Results and discussion

The developed models are solved numerically and parametric studies are carried out. The variations of gas compositions with equivalence ratio in the gasification zone are shown in **Figure 5**. The result depicts a slight decrease in the percentage composition of CH_4 with an increase in the equivalence ratio. The decrease is simply attributed to increased moles of air intake into the gasifier. It could be observed that the variation of H_2 concentration as the equivalence ratio increases follows an inverse trend. This is because higher availability of O_2 first consumes H_2, which is also reflected in an increasing concentration of H_2O. The same trend was also observed in the variation of CO with the equivalence ratio. The decrease in CO concentration as the equivalent ratio increases is due to oxidation of CO at higher equivalence ratio, which is further validated by the increasing trend of CO_2 concentration with equivalence ratio. It could therefore be stated that the compositions and distributions of CO, H_2 and CH_4 in the gasifier decrease as the equivalence ratio increases. However, the temperature distribution in the reactor increases with an increase in the equivalence ratio.

In this work, the performance of the biomass gasifier system is evaluated through the equivalence ratio, the syngas composition, cold gas efficiency and lower heating value (LHV). **Figure 6** presents the variation of gasifier performance parameters i.e., cold gas efficiency, gas yield and LHV of gas with equivalence ratio. As it could be seen in the figure that the cold efficiency and LHV decrease as the equivalence ratio increases while the gas yield increases with an increase in the equivalence ratio. The increased gas yield is due to higher air intake as the equivalence ratio increases while the decrease in cold gas efficiency and LHV of gas may be attributed to the consumption of combustible gas due to more availability of air as the equivalence ratio increases.

Figure 7 shows the impact of oxygen consumed on gas produced while **Figure 8** depicts the influence of gasification zone temperature on the CO/CO_2 ratio in the gasification zone. From the **Figure 7**, it is shown that the quantity of gas produced

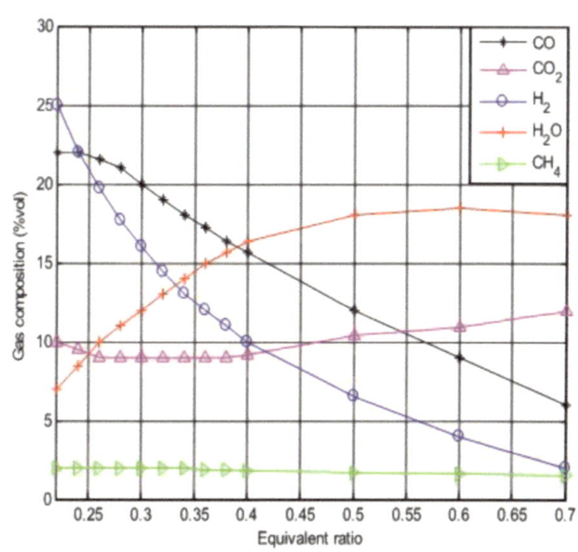

Figure 5.
Effects of equivalent ratio on the percentage gas composition.

increases as the amount of oxygen consumed increases. This is due to increasing in the rate of combustion of the products of pyrolysis as more oxygen is consumed in the combustion process or oxidation reactions, more gases are produced from the reactor. It is also shown in **Figure 8** that the ratio of CO/CO_2 decreases as the temperature of the reduction zone increases.

The validation of the computational fluid dynamic simulations is very necessary. The comparison of the computational fluid dynamic simulations with the results of the experiment [33, 34] as shown in **Figure 9**. The results showed good agreements with the results of the measurement.

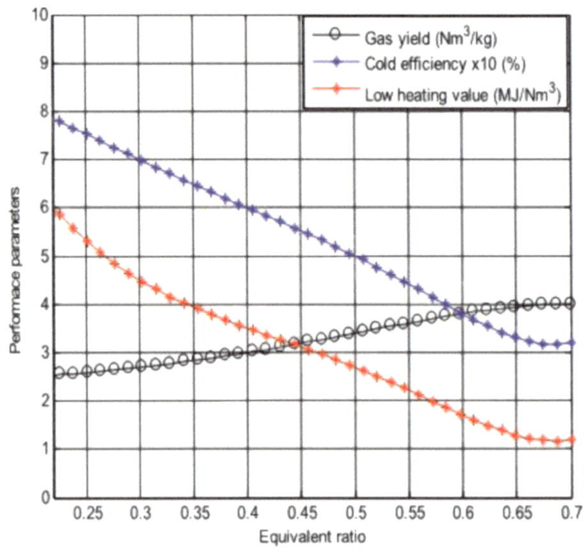

Figure 6.
Effects of char particle diameter on the char combustion rate.

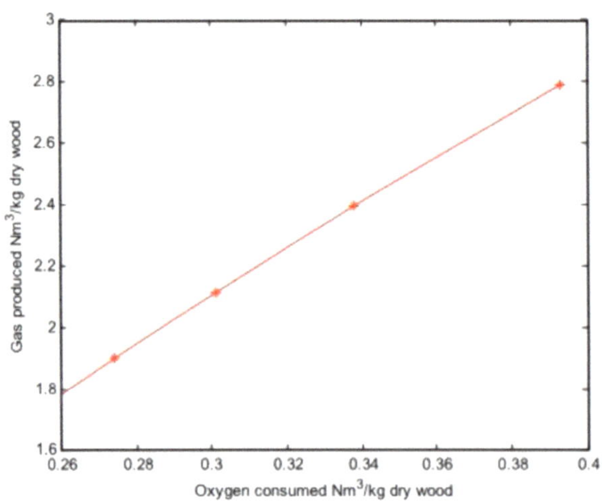

Figure 7.
Effects of oxygen consumed on gas produced.

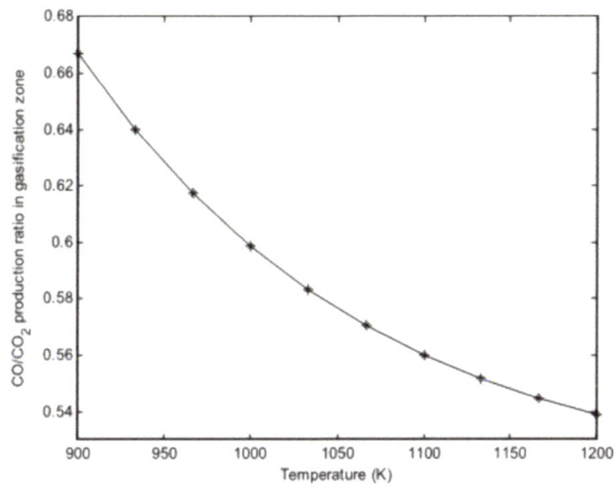

Figure 8.
Effects of temperature on CO/CO_2 ratio in the gasification zone.

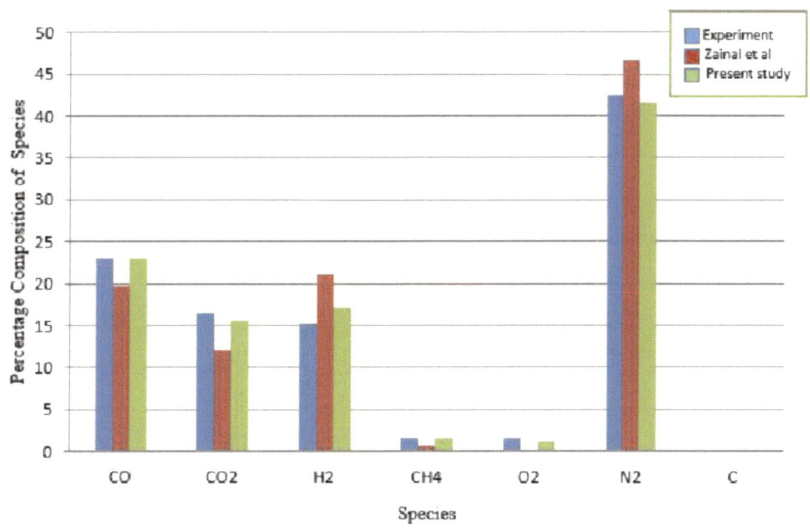

Figure 9.
Comparison of the results.

5. Conclusion

In this work, improved kinetic models for the combustion and gasification zones in the thermochemical system have been developed. The performance of the biomass gasifier system was evaluated through the equivalence ratio, the syngas composition, cold gas efficiency and lower heating value. Also, the effects of the equivalent ratio on gas compositions, the gasifier performance and the low heating value of the biomass were investigated. From the analysis, it is established that

i. the concentration of CO, H_2 and CH_4 in the gasifier decrease as equivalence ratio increases.

ii. the CO_2 concentration increases with increase in the equivalence ratio.

iii. the cold efficiency and LHV decreases as the equivalence ratio increases while the gas yield increases with increase in the equivalence ratio.

iv. the quantity of gas produced increases as the amount of oxygen consumed increases.

v. the ratio of CO/CO_2 decreases as the temperature of the reduction zone increases. The developed model is validated with the experimental results.

The analysis is far less costly and time consuming than the experimental approach. Such analysis is useful as a time saving and cost effective tool for designing and optimizing the biomass gasifier.

Author details

Sunday J. Ojolo* and Musbau G. Sobamowo
Department of Mechanical Engineering, University of Lagos, Nigeria

*Address all correspondence to: sojolo@unilag.edu.ng

References

[1] L. Devi, K. J. Ptasinski, F. J. J. G. Jenssen. A review of the primary measures for tar elimination in biomass gasification processes. Biomass Bioenergy, 24(2003), 125–140.

[2] P. Quaak, H. Knoef and H. Stassen. Energy from Biomass: A Review of Combustion and Gasification Technologies. Energy Series World Bank Technica'l Paper NO. 422, I44, 1999.

[3] D. L. Klass. Biomass for Renewable Energy, Fuels, and Chemicals. Academic Press, San Diego, 1998, 1-651.

[4] S.Yang., H. Wang, Y. Wei, J. Hu and J. W. Chew. Eulerian-Lagrangian simula tion of air-steam biomass gasification in a three-dimensional bubbling fluidized gasifier," Energy, Elsevier, 181(C), (2019), 1075-1093.

[5] A. Klein. Gasification: An Alternative Process for Energy Recovery and Disposal of Municipal Solid Wastes, 2002.

[6] A. Klein and N. J. Themelis. Energy Recovery from Municipal Solid Wastes by Gasification. North American Waste to Energy Conference (NAWTEC 11)11 Proceedings, ASME International, Tampa FL (April 2003), 241-252, 2003

[7] D. F. Fletcher, D. F. Haynes, B. S. Christo, F. C. Joseph, S. D. A CFD based combustion model of an entrained flow biomass gasifier. Applied Mathematical Modelling, 24(3)(2000), 165- 182.

[8] L. Gerun, M. Paraschiv, R. Vijeu, J. Bellettre, M. Tazerout, B. Gøbel, U. Henriksen. Numerical investigation of the partial oxidation in a two-stage downdraft gasifier. Fuel, 87(2008), 1383 – 1393.

[9] M. Zhou, L. Yan, Q. Guo. Non premixed combustion model of fluidized bed biomass gasifier for hydrogen rich gas. Chinese Journal of Chemical Physics, 19, 2006.

[10] K. Papadikis, A. V. Bidgwater and S. Gu. CFD modelling of the fast pyrolysis of biomass in fluidized bed reactors", Part A: Eulerian computation of momentum transport in bubbling fluidized beds", Chemical Engineering Science, 63, 4218 - 4227, 2008.

[11] Y. Wang and L. Yan. CFD studies on biomass thermo chemical conversion", Int J Mol Sci, 9, 1108 – 1130, 2008

[12] H. Yilmaz, S. Perumalsamy, G. Oeljeklaus, K. Gorner, T. Klasen, A. C. Benim. CFD Analysis of a Large Scale Entrained-Flow Gasifier for IGCC Power Plants. 7th International Conference on Multiphase Flow. Tampa, FL: ICMF, 2010.

[13] P. Cornejo and O. Farias. Mathematical Modeling of Coal Gasification in a Fluidized Bed Reactor Using a Eulerian Granular Description. International Journal of Chemical Reactor Engineering, 9(1)(2011), 1-30.

[14] T. Paes. Modeling for control of a biomass gasifier. Eidhoven: Technische Universiteit Eindhoven, 2005.

[15] H. Watanabe and M. Otaka. Numerical simulation of coal gasification in entrained flow coal gasifier. Fuel, 85(12-13). 2006, 1935-1943.

[16] H. J. Huang and S. Ramaswamy. Modeling Biomass Gasification Using Thermodynamic Equilibrium Approach. Applied Biomechical Biotechnology, 154 (1-3) (2009), 193-204

[17] A. Cuoci, T. Garavelli, A, Frassoldati., R. Grana, S. Pierucci, E. Ranzi and S. Sommariva. Mathematical modeling of gasification and

combustion of solid fuels and wastes. Chemical Engineering Transactions, 18 (2009), 989-994.

[18] S. M. Atnaw and S. A. Sulaiman. Modeling and simulation study of downdraft gasifier using oil-palm fronds. Conference: 3rd International Conference on Energy and Environment. ICEE 2009.

[19] M. Hamzehei, H. Rahimzadeh, G. Ahmadi. Studies of gas velocity and particles size effects on fluidized bed hydrodynamics with CFD modeling and experimental investigation. Journal of Mechanics; 26(3), 2010.

[20] L. Tingwen, G. Aytekin, S. Madhava. High-Resolution simulations of coal injection in a gasifier. Ind. Eng. Chem. Res., 49(2010), 10767–10779.

[21] H. Zhang, S. Shao, R. Xiao, Q. Pan, R. Chen, J. Zhang. Numerical study on the hydrodynamics of a self-heating biomass fast pyrolysis reactor" Energy Fuels, 25(2011), 4077 – 4084, 2011.

[22] B. Peng, J. Zhu and C. Zhang. A new approach to specify the inlet boundary conditions for computational fluid dynamics (CFD) modeling of hydrodynamic behavior of a circulating fluidized bed (CFB) riser. Ind. Eng. Chem. Res., 51(2012), 2152 – 2165.

[23] W. Chen, X. Liu, X. Fan, L. Chu, Y. Yang, W. Liu. Three-Dimensional Simulation of Gas-Solid Flow in the Biomass Circulating Fluidized Bed Gasifier's Riser, Advanced Materials Research, 383-390(2012),6537-6542.

[24] Y. Wu, Q. Zhang, W. Yang, and W. Blasiak, Two-Dimensional Computational Fluid Dynamics Simulation of Biomass Gasification in a Downdraft Fixed-Bed Gasifier with Highly Preheated Air and Steam. Energy & Fuels, 27(6) (2013), 3274-3282.

[25] H. Luo, W. Lin, W. Song, S. Li, D-J. Songgeng, H. Wu. Three-dimensional full-loop CFD simulation of hydrodynamics in a pilot-scale dual fluidized-bed system for biomass gasification. Fuel Processing Technolog y, 195(2019), 106146.

[26] H. Liu, R. J. Cattolica, R. Seiser, C. Liao. Three-dimensional full-loop simulation of a dual fluidized-bed biomass gasifier. Applied Energy, 160 (2015), 489-501

[27] D. Lu, K. Yoshikawa, T. M. Ismail and M. A. El-Salam. Computational Fluid Dynamics Model on Updraft Gasifier using Carbonized Woody Briquette as Fuel. 9th International Conference on Applied Energy, ICAE2017, 21-24 August 2017, Cardiff, UK

[28] U. Kumar, A. M. Salem, M. C. Paul. Investigating the thermochemical conversion of biomass in a downdraft gasifier with a volatile break-up approach. 9th International Conference on Applied Energy, ICAE2017, 21-24 August 2017, Cardiff, UK

[29] K. G. Santos, V. V. Murata and M. A. S. Barrozo. Three-dimensional computational fluid dynamics modelling of spouted bed. The Canadian Journal of Chemical Engineering. 87, 211-219, 2009.

[30] J. K. Kone, X. Zhang, Y. Yan, G. Hu, and G. Ahmadi. Three-dimensional multiphase flow computational fluid dynamics models for proton exchange membrane fuel cell: A theoretical development. The Journal of Computational Multiphase Flows, 9(1 (2017), 3–25

[31] Biomass Engineering Limited, Newton-le-Willows, Warrington, UK

[32] Jain, Ankur Scientific, Baroda, Indian March 2006 and Senior Capstone Design, Villanova University, 2006.

[33] Z. A. Zainal, A. Rafiu, G. A. Quadir and K. N. Seetharamu. Experimental

investigation of a downdraft biomass gasifier. Biomass and Bioenergy 23(4) (2002), 283-289.

[34] Z. A. Zainal, R. Aliu, C. H. Lean and K. N. Seetharamu. Prediction of performance of a downdraft gasifier using equilibrium modeling for different biomass material. Energy Conversion and Management 42(12) (2001):1499-1515.

Chapter 8

Design and Evaluation of an Automated Monitoring and Control System for Greenhouse Crop Production

Arasu Sivagami, Michael Angelo Kandavalli and Bhaskarrao Yakkala

Abstract

An embedded system integrated with sensors based on nanomaterial is proposed for closely monitoring and control microclimate parameters 24 hours a day to maximise production over the whole crop growth season by introducing greenhouse for the cultivation of plants or specific plant species. The system will also eliminate errors in human intervention to optimise production of crops. This system consists of sensors and actuators, an Analogue to Digital Converter (ADC) and a Raspberry Pi. The system will determine whether a defined threshold is passed by any climate parameter and systematically changes via the controller. The current work reduces human input through automated irrigation to optimally utilize a scarce resource, namely water. Climatic parameters for plant growth such as, moisture, humidity, temperature, water pressure in drip pipe, soil salinity etc. are monitored and optimized. Furthermore, work was extended to include GSM to control the entire farm remotely. For its success, it is very important to choose a greenhouse location. For instance, the problems are quite different when choosing an adjoining greenhouse, for instance a sunroom or greenhouse. The greenhouse location should be chosen for sunlight, proximity to power and water sources, wind, drain and freeze pockets, and the proximity of the garden and house. The intention behind accomplishment and devise of GSM based Fertigation System is to construct and evaluate the requirement of water in the yield as farming is the major resource of production which habitually depends on the water accessibility. Irrigation of water is usually done by manual method. To ease the work of the farmer GSM based automatic Fertigation (includes chemigation too) system can be implemented so that water wastage can be reduced and also the fertilizer can be added accordingly. Also the Soil Salinity can be checked and reduced if exceeds certain limit. By using GSM, only GSM command via GSM mobile can control the start and stop action of a motor that feeds the field with the water. GSM is used for controlling the entire process and the entire system backbone. It can be used from any distance to control irrigation. The results are assessed by electronic simulator PROTEUS using the desired optimised parameters, the design of this automated greenhouse system with PIC controller. As the inputs to the microcontroller and as an LCD screen record the respective outputs, the model produces a soil moisture sensor, light sensor and temperature sensor. The system performance is accurate and repeatable for measuring and controlling the four parameters that are crucial for plant growth - temperature, humidity, soil moisture and light intensity. With the reduction in electricity consumption, maintenance and complexity, and a flexible and precise environment control form for agriculture, the new system successfully cured quite a couple of defects in existing systems. Nano composite film sensors (Graphene and Graphene mixed in order to optimise the input of fertilisers for chemical composition determination. Using nano technology in agriculture enforces the firm bond between the engineer and farmer. Nano material film-based gas sensors were used to measure the presence of oxygen and CO2.using graphene nano composite sensors integrated into an embedded system, to detect the presence and levels of gases. Improve crop growth with combined red and blue light for lighting under the leavened and solar-powered LED lighting modules. This was achieved by graph/solar cells. The light was measured at the photosynthesis flux (PPFD) of 165 μmol m-2 s-1 by 10 cm of its LED module. LED lights were provided between 4:00 a.m. and 4:00 p.m. in the daytime treatments and night treatments from 10 to 10 hours. The use of the nighttime interlumination of LEDs was also economical than the interlumination of charts. Thus, nightlighting LEDs can effectively improve plant growth and output with less energy than the summer and winter times. Solar panels are best functioning during times of strong sunlight today, but begin to wan when they become too hot and cloudy. By allowing Solar Panels to produce electricity during harsh weather conditions and increase efficiency, a breakthrough in graphene-based solar panels can change everything. Ultimately with a fully autonomous system, agricultural productivity and efficiency, the length of the growing season, energy consumption and water consumption were recorded and monitored by exporting the data over GSM environment. With the steady decrease in the cost of high-performing hardware and software, the increased acceptance of self-employed farming systems, and the emerging agricultural system industry, the results will be reliable control systems covering various aspects of quality and production quantity.

Keywords: Nano sensors, Grapheme material based Solar cells LED, supplemental lighting, lighting period, photosynthesis, yield, Fertigation, Chemigation

1. Introduction to automated Irrigation

1.1 Basic concepts of automation

A device containing inbuilt program that performs governing or controlling a flow of water from one zone to another zone absence of the irrigator [1, 2].

Mechanization can be utilized in various manners:

- to start and stop irrigation through fluid channel outlets,
- to start and stop tube,
- to remove the progression of water from one water system zone – either a straight or a segment of channel and guiding the water to another zone.

This improvement provides without any direct manual effort, but the irrigator may need to spend time preparing the system at the start of the irrigation and maintaining the components, so it works properly.

1.1.1 Merits and demerits of automatic irrigation

Merits:

Reduced labour: As a water supply is not necessary to continuously screen the progress of a water system, the water supply system is available to carry out various tasks on an ongoing basis.

Improved lifestyle: The irrigator is not necessary for the water to be dampened uniformly downward. The irrigator can stay away from the assets and sleep all night long with the family.

Faster irrigation: robotic irrigators prefer to irrigate if water is needed by the plants, not when it suits the irrigation system.

Helps with higher flow rates: Many irrigators aim to increase the watering rate by installing higher channels and narrow outlets. Such flow rates usually require more work as a bay's time to water is reduced and therefore more and more change is required. Robotization admits to be handled without an increase in the number of work for this higher procedure.

More precise cut-off: The water system automation allows water cut-off at the exact narrow point. Ultimately, this is more accurate than manual inspections, since there may be errors if the operator's water flow changes too late or too early.

Reduced water and nutrient overflow: automation can help maintain fertiliser on farm by reducing efficiency in landfill. The retention of fertiliser on farm benefits both economically and environmentally [3].

Demerits:

Cost:There are costs in buying, installing, and maintaining automatic equipment.

Reliability:Can the irrigator trust a programmed framework to work accurately unfailingly? Now and again disappointments will happen. Frequently these disappointments are a direct result of human blunder in setting and keeping up the frameworks. A re-use framework is acceptable protection to gather any overabundance spillover when disappointments happen.

Development of channel preservation:There is a need to expand upkeep of channels and hardware to guarantee the framework works accurately. Channels ought to be fenced to shield the programmed units from stock harm.

1.2 Automated watering systems types available

Air system: Pneumatic: A pneutic frame, initiated by a narrow sensor at the cut-off point, is an unchangeable framework. The air channelled to the instrument for opening and shutting water system structures is then pressurised when water enters the sensor [4].

Compact system of timers: A multi-faceted clock frame is an impermanent system that uses electronic tickers to open and shut down the water system structures. Due to its compact nature, property owners usually purchase 4 or 5 units to move around.

Hybrid based timer/sensor: This is a crossbred of compact clock and sensor frames, as the name suggests. It uses an electronic gadget to open and shut off structures for the water system, like a convenient clock. This frame has in any event an additional component of the river, which can be used to direct a moveable sensor down that transmits radio signals to the clock gadgets at the outlets for opening or closing the structures when it interacts with water, and sends a radio message to the collector saying that the landlord has come to the chopped water centres.

SCADA: the Supervisory Control and Data Acquirement (SCADA) use robotic frameworks consist of a PC and a programme bundle with timing and controlling a radio system for water. Signs from the PC are sent to the enclosure to control the water system structures with straight actuators and open and close them. Straights are opened and shut down on a premise; a few frameworks can naturally change the time when the channel is in conflict when a sound outlet is opened. The extra advantage of SCADA-based frames is that they are able to start and stop water pumps and motors.

1.2.1 System selection

All mechanisation structures have focal points and burdens to be viewed as to what framework is appropriate for a particular property to the water system format. No framework is the best for all properties. There is no framework.

The methods used by the irrigator for the water system should be considered. Should a framework be required which is able to move around the property and possibly used on various properties, the irrigator must consider the versatile frameworks in this case. If the irrigator requires a frame where the components are fixed and a similar water system can be used which group each water system, a fixed framework would gradually be appropriate in such a case.

The irrigator should consider the expenses of the frame when deciding the best framework for an estate, make sure the framework is adjusted and which framework is most appropriate to the property and the water system [4].

The objectives to consider are:

1. Disentangle the water system framework by introducing and structuring the entire water system framework.

2. Spare vitality which permits the use of savvy water system framework.

3. Advance water utilization.

4. Mechanized framework completely.

5. Decrease the intercession of human.

6. Make framework simple to use by ranchers.

1.2.2 History of automated irrigation

By definition, the water system is the function that water dries through tubes, which contribute to the cultivation of the plant and plants. In your own yard, it's easy to honour and the aerosol you run. After having a settled grass spray system, you won't have to return to your own pants. Attentive experts are guaranteed the repair and installation of water systems that satisfy your requirements and accreditation by aerosol or sprayer [5].

It produced widespread interest during the 19th century when European Americans flooded the drought-prone plains. A water system was practised when a moat was brought to gardens, lawns and trees by a particular area in the 1870's called Fort Sidney. Until 1890, when a drought began, public opinion for the water system was not as conducive then to irrigation. This strengthening of irrigation began to increase every day; there were conventions and legislation was even proposed in the United States. But a while later, Farmers soon appeared to be irrigated by the floods. Sodium and calcium carbonates were rapidly transferred in the land, which makes it too salty for the development of convent plants [6].

The 1930s, also known as the Dirty Thirties, were an era of inflexibility and cultivation of poor humidity in the region, which generated soil arid habitually. Farmers moved quickly to irrigation. The achievement of sprayer systems throughout Canada has been enhanced by modern technologies. Flood irrigation is still used; however, it is no longer responsible for tending crops, normally used only in the sugar industry. It has no responsibility. Pipes and sprayer systems are new and more modern uses of the water system [6].

Turf Rain Irrigation Systems provides over 24 years to the region of South Ontario including Toronto, Mississauga, Oakville, Milton, Burlington, Brampton and neighbouring GTA. We are specialist in pond sprayers, water systems, residential, business, industrial and golf landscape lighting. They also maintain the rehabilitation and maintenance of irrigation systems [5].

1.2.3 Mechanized water system using solar power in Bangladesh

The gadget specializes in rice fields in nations depending on agriculture within theeconomy, such as Bangladesh. The primary concept in this gadget is to cognizance on the level ofwater in agricultural fields because those fields lose lots of their merchandise due to floods. Thesensor sends a message from the field to the person approximately the extent of water within thearea if it will increase or decreases then the operator controls the pump to regulate or flip off thetelephone. The blessings of this machine are that it depends on the sun energy to get hold ofelectricity. The dangers of this system are that it centered on one sort of sensor, the water stagesensor, no matter whether the plant desires water or not. There may be no opportunity source ofenergy in case there is no solar electricity to run the device.

1.2.4 Construction and implementation of a mechanized water system in Nigeria

In this machine the basic idea is to rely on the type of soil and the amount of water needed by each type of soil. This process is done by measuring the level of moisture in each type and usingthe pump to supply water. The result indicates that sandy soil requires less water than clay soils.The consents of this device are to focus on soil moisture and water conservation. But making themachine much less powerful is to measure the moisture of soil from one location in the agriculturalland. It is far viable that the vegetation at the other end of the rural land does no longer needwatering. Also, the water source is not constant.

1.2.5 Solar water system

In many of the development projects, the admission of solar pumps to drylands like Africa, India and South America also shows the increase in the ability of local farmers to improve their living conditions. One of the good examples of this is that the scientists developed two solar pumps to support Wedel SET GmbH in an attempt to teach physics at a Blankenese School in Germany. These underground water pumping systems have been installed on two farms in Nicaragua. In addition the UNAN University in León, who worked hard to harness solar energy, was able to achieve this project. This project has been in operation since over 10 years and in Nicaragua now there are 30 pumps. The Nicaraguan company Enicalsa is under the supervision of farmers who benefit from the solar water system. Using solar pumps, even in dry seasons, solar pumps can generate year-round, thus increasing their revenues and reinforcing their local market conditions [7].

In addition, there is an increase in interest in European solar water systems beyond the already noticed regions. The production phase from Austria reached a few months ago a mobile solar drip water system. This project was carried out by the Austrian company Wien Energie with a dual goal: To reduce CO2 radiation, on one hand, using solar power, and to achieve 30 per cent water savings compared to conventional water sprinklers with the dripper irrigation method, on the other hand.

It is quite simple to assume the drip water system method. With different brakes, tubes and pipes the water slowly and at normal intervals is drained into the roots of the plants. There is no wasted water, because water is legally sprayed in the air, or leaked into soils where plants do not grow, unlike a spraying system. The dripping water system method therefore allows more crops to grow with less water, making it a highly productive water system process.

Wien Energy solar water systems are connected to a wheel pump that can power and connect to a smartphone application and can determine the energy generated by the system. The mobile photovoltaic system with up to 3kw is combined with a wheelable pump. The solar pump then allowed the water directly to the plants via the tubing. This system is now ready to be produced following a thorough test of 3.5 hectares of organic maize in Guntramsdorf, Austria [8].

Thus, the drinking water system can contribute to efficient water management in countries with high temperatures and insufficient water resources. This is important because farmers are faced with three challenges: water, money and energy savings. Mobile solar drip water systems are the perfect answer to these problems. While these systems are costly and complicated to resolve, many research and development projects aim to make the use of sun power democratic in agriculture, which in future can play a vital role in the management of food and energy crises. (and even now).

A system with an RFID-based wireless sensor network is proposed in this article [9] by the author. Within this system, the author sets moisture on the ground at different places in the field, i.e. farm or farm. Now the sensor transmits data at 2.45 GHz to ZigBee. Now the sensor sends these data to the base plant, and when the soil is dry then only this part of the field is fed by the pump plant [10].

The smart irrigation system based MQTT protocol is produced by Ravi Kishore Kodali and.al. The moisture sensors, soil sensor and water pump are used by Esp8266 Node MCU-12E. This system is used to transfer the data between Esp8266 NodeMCU-12E and the sensor by using the QueueTelemetry Transport Protocol (MQTT) Message system. Soil moisture transmits data to Esp8266 MCU-12E Node and, if the soil is dry, Esp8266 NodeMCU-12E is sent instructions to pump water and to pump the water. They display the actual soil and water pump status using LCDs. Sneha Angal constructs an office and home planting system. This system is equipped with a sensor for raspberry, Arduino, ZigBee and soil humidity. The main control block is in the proposed raspberry pi system. The instructions sent from Arduino are processed. The sensor of moisture is attached to Arduino and ZigBee is a middle between Arduino and raspberry pi. This system is modular, so if any module does not work, the user can modify it. To improve this system, the GSM module can be added to achieve soil status and watering plant by miscalculating the number of the GSM module [10].

Mare Srbinovska et.al developed a smart, vegetable greenhouse wireless sensor network. Humidity, temperature and lighting are being measured in this system. The first steps of this system are to measure the data transfer capacity and select the data exchange algorithm. The second phase of the system proposed determines the design and development of the system based on experimental findings. The last step is to test, analyse and optimise the wireless sensor network.

They focus in this paper [8] on very important agricultural products issues. They created a system to detect tomato disease. They made a robot to continuously shoot a plant. Now they have produced a video processing algorithm afterwards. The first phase focuses heavily on the classification of the tomato plant, while the second is the recognition of diseases at the border of tomatoes. They use the k-mean clustering algorithm to identify diseases.

Lala Bhaskar et.al are constructing a system to increase food quality and productivity. Different factors such as temperature, humidity, water level of the ground and LCD are measured by this system. In order to inform farmers of the current status via the SIM900 module about the registered number, the data is monitored and sent to farmer with the message. Sensors such as soil moisture sensor and temperature sensor are used. This system is helpful for farmers with a power failure and uniform water distribution due to electricity failure.

This paper [11] aimed to build a cheap system because it can be used by an anonymous farmer, and a step towards intelligent agriculture. Raspberry-pi, soil humidity sensor and GSM module were used in this system. If soil humidity is sensed as dry soil, the farmer will be notified and mailed to the registered email address. Local Shortest Path (LSP) was used by the proposed system author in order to control the wireless sensor network, i.e. to obtain data from the sensor via LSP.

They proposed a system to improve the agriculture method, to measure the PH rate to detect dryness of the soil and also to keep an eye on temperature and water level. The pi, LCD, and soil humidity sensors were used to show current status and GSM module. Now that a certain threshold value decreases the PH rate, it notifies the user of improving the method of agriculture and proposes to farmers based on the PH value. They used LM35 for soil temperature measurement. They continue to collect data in this system to find water levels to enhance the system by watering the ground on a water level basis.

Ch Sumaliya et.al suggested a low cost system. The controller, soil moisture sensor and temperature sensor ATMEGA328 was used in this system. The data was displayed in LCD using a raspberries pi and ZigBee recipient. Now, if soil moisture falls from a particular limit value, the buzzer will be littered and the state on LCD is displayed. When the threshold moisture increases, they start the motor and the buzzer disappears. They don't use Wi-Fi in this system so that we can build a Wi-Fi system. The temperature sensor can also be added and instructed on the motor for measuring the temperature.

The focus of this article [6] was a saving water by means of an intelligent irrigation system. They focus primarily on gardens, plants and parks to automatically supply water. And water supply based on requirements requires additional water and some require less water. They use a microcontroller to ensure the requirements and obtain data from the sensor of soil moisture and temperature. If the water level is low, plants can receive water to improve this system.

Bin Bahrudin, Md Saifudaullah and.al proposed stayed for fire. The raspberry pi, Arduino Uno and the smokesensor, camera module and the GSM module have been used on this system. The system clicked a photo and shows it on the website when smoke is detected. It now requires users to confirm that there is or is not a fire. In the event of a fire, the fire task force sends SMS. This system is now used to reduce the amount of storage needed to store image to counterfeit because the camera clicks on the photo.

In this paper [12] authors developed a system for intelligent irrigation and weather surveillance. To this end, they take into account certain parameters such as soil humidity, humidity, temperature, rainfall, and wind direction. They measured soil humidity to detect soil dryness and rain sensor soil evaporation. So, if the moisture level of the soil falls from the user value, the watering starts. For mapping wind speed, a device called anemometer. This data is uploaded to the server and shows the data on the LCD.

2. Introduction to nano sensors

The nano sensors are nano sensors that calculate and switch physical quantities to identify and inspect the signals. Nano sensors Today we have a number of approaches for the production of nano-sensors: top-down lithography, bottom-up and molecular self-montage. A variety of nano sensors, particularly in the defence, ecological and healthcare sectors, is available on the market. The same primary workflow is distributed by these sensors: a permanent selective analyser, signal generation from the nanosensor relationship to its bioelement and signal generation in useful metrics [13].

Nanomatherapy sensors differ from sensors made from traditional materials because they do not appear in a mass material occurring on the nano-scale, so they have different sensitivities and specificities. Nano sensors operate on the same scale as natural biological processes, allowing for noticeable physical changes with chemical and biological molecular functionally. Sensitivity amplification is caused by the high ratio between the surface and volume of the nanoparticles and the physical narrational properties of nanomaterials, including nanophotonics, that can be used to detect them. Nano sensors are used to add fundamental processing capabilities for the nano sensor together with nanoelectronics [13].

In addition of sensitivity and particularity to the nano sensors, it offers a valid advantage in cost and time response, its accomplishing for high-throughput applications [14]. Nano sensors accommodate real-time monitoring compared to traditional detection methods such as chromatography and spectroscopy. The traditional methods may take more to access the results and often lack of asset in capital costs together with time for sample preparation.

One-dimensional nano materials like nano wire, nano tube are well suited for use of nano sensors, in comparison with bulk or thin –film planner devices. That can be worked as transducers and wires to transmit the signal. Their high surface area can cause large signal changes upon binding of an analyte Their small size can enable large scale multiplexing of severallycapable of addressing sensor units in a small device,their operation is also "label free", is not essential fluorescent or radioactive labels on the analytes. Zinc oxide nanowire is employed for gas sensing applications, given that it demonstrates high sensitivity towards low concentration of gas beneath ambient conditions and can be manufactured easily with low cost.

By avoiding drift and fouling, developing reproducible position methods, applying pre concentration and departure methods to get a proper analyte combination that avoids overload, and accommodate the nano sensor with other sensor elements package in a stable manufacturable manner. After all nano sensors are almost new technology, there are lot of unanswered questions related to nanotoxicology, which presently permits their application in biological systems.

Possible nano sensor applications include medicine, contamination and pathogens detection and monitoring of production and transport systems. At the molecular level, nanosensors can differentiate between certain cells and can either carry out medicines or monitor their development of individual body locations by calibrating physical properties change (volume, concentration, displacement and velocity, force, electrical and magnetic forces, pressure and temperature). The signal transduction type defines the main nano-sensor allocation system. Various types of read outs for a nano-sensor include optical, electromagnetic, mechanical or vibrational. Nanosensors that use molecular imprinted polymers (MIP) can be divided into 3 classes of electrochemical, piezoelectric or spectroscopic sensors as an example of categorization. Electrochemical sensors induce a change in sensing material's electrochemical properties including charging, conductivity, and electrical potential. Piezoelectric sensors convert mechanical energy into electrical

DOI: http://dx.doi.org/10.5772/intechopen.97316

strength or vice versa. Then this force is transferred to a signal. MIP spectroscopic sensors, such as chemical light, surface plasmon resonance and fluorescence sensors, can also be divided into three sub-categories. These sensors produce light-based signals in the forms of chemical light, resonance and fluorescence, as their name suggests [13].

2.1 Embedded system design with nano interpretation

The following **Figure 1** shows the simple drip irrigation model and **Figure 2** describes the corresponding embedded transformation.

The structure is an embedded system that regularly examines and controls the microclimatic parameter of a field for crop or species of specific plants, allowing production to maximise in the entire crop season and removing difficulty. The main purpose of this system is to improve and respond effectively and reduce human interference, which also increases levels of protection. It includes sensors, microcontrollers, analogue to digital converters and actuators. When any climate parameter crossing a security threshold to protect the crops is kept, the sensors feel changes. After converting the ADC to a digital form it is read on the data from the input ports by the microcontroller. The microcontroller takes action by using relays up to the optimum level of the stretched out parameter. As the core of the system is a microcontroller, the setup is low cost and yet efficient. The entire setup is affable because the system also uses an LCD display for constantly alerting the user of the circumstances in the field [9].

The irrigation system requires approximately half of a sprinkler or surface irrigation system's amount of water. Lower operating pressures and flow rates lead to lower energy loads. You can achieve a higher degree of water management [15]. More exact amounts of water may be provided for plants. The damage to insects and diseases is reduced by keeping leafy plants dry. Investments generally decrease in

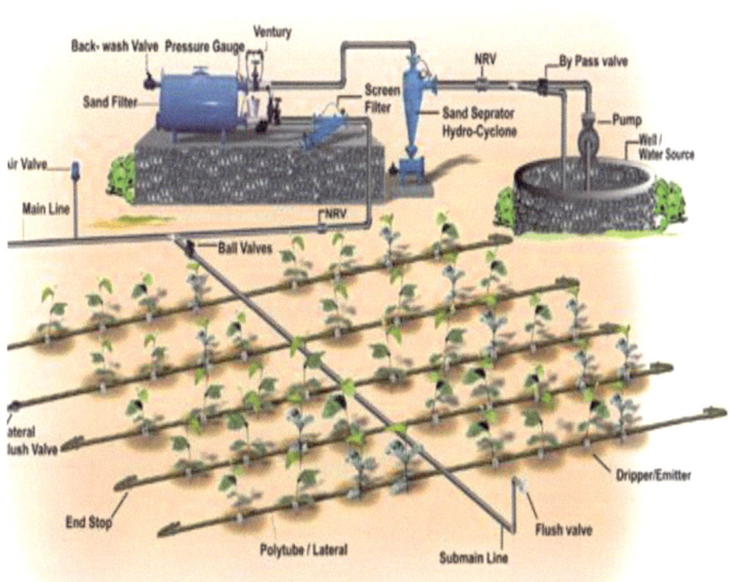

Figure 1.
Simple Drip irrigation Model.

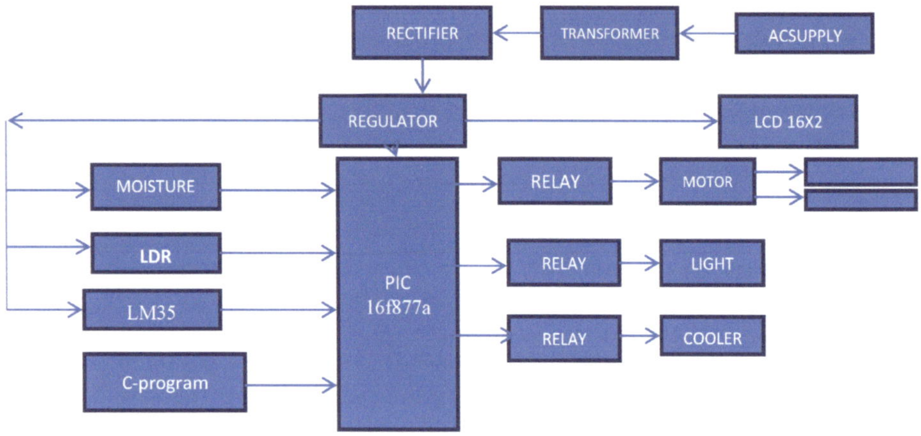

Figure 2.
Embedded System Transformation.

operating activities. Due to dryness of rows between plants, the process can continue during irrigation [16].

A drip irrigation system may be used under a wide range of field conditions. A typical drip irrigation system is displayed [10]. Drip irrigation is trendy, as it can increase yields and reduce requirements for water and labour.

This type of system can apply manures. This can lead to lower dung and costs. Drip watering results in less soil and a wind erosion if it differentiates with overhead sprinklers [17].

2.2 Smart irrigation using Raspberry PI and GSM

With the Raspberry pi, GSM module, soil humidity sensor, flame detector, ultra-sonic sensor, and buzzer, the above embedded system could be changed. The Raspberry Pi is only a computer and has a very powerful, lightweight ARM processor. It also has USB ports, WiFi modules, HDMI port and Ethernet port, Raspberry Pi 3 Model B. OSes such as Raspbian, Ubuntu MATE, Snappy Ubuntu, Pidora, Linutop, SARPi, Arch Linux ARM, Gentoo Linux, freeBSd, KaliLinux and RISC OS Pi are also available for Raspberry Pi [18, 19].

It has multimedia application support as a small computer. The reason is HDMI and the support for graphics. But it does also have some limitation, but we can insert a micro SD card, so we can boot the Raspberry Pi OS. But there are no limitations.

We use a raspberry pi, soil moisture sensor, a flame sensor, an ultrasound sensor, a buzzer and a servo engine in this system. The heart of the system is the raspberry pi, i.e. the main system control unit. We used raspberry pi b+, which has many new characteristics in this system. It has also improved IO connectivity compared to the older pi version. Soil humidity is connected directly with raspberry pi. Now when the sensing data is transferred, the data is transferred to raspberry pi [20]. Raspberry pi reacts according to the soil humidity sensor received data. If soil is dry, send an email to the mobile number and email address registered with the farmer and the motor start. The detection of fire in a farm is used to detect fire. The fire sensor collects data from the field and passes it on to the pi. If fire is on, we will now send a message to the farmers' registered mobile telephone and e-mail address. We simultaneously blow up the buzzer. We measure water level well with an ultra-sonic sensor. We will now take the depth and radius of the well of farmers to measure water levels in a good way. Now we can measure the current water level in the well by using an ultrasonic sensor and calculate the water level from the bottom of the well. We will subsequently send it to the registered mobile and email address of the farmer [10].

Figure 3.
Raspberry pi.

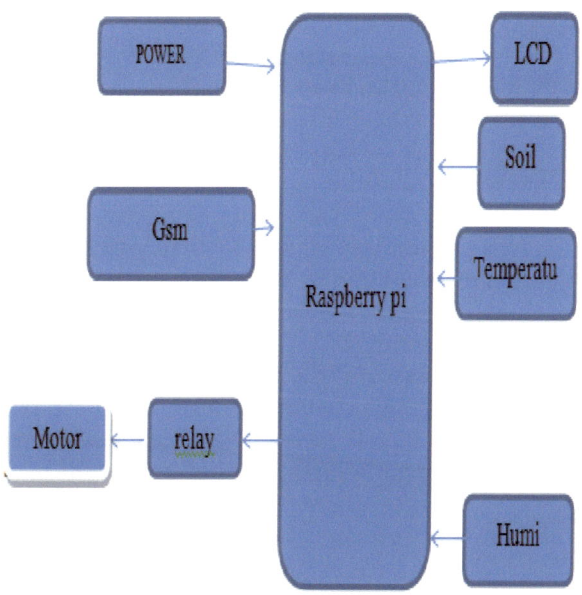

Figure 4.
Smart Irrigation with Raspberry Pi and GSM.

Film-based nanomaterial enhanced polymer sensors (Graphene and Graphene mixed to form a definite chemical soil structure for improved input manure. Thus, through these systems, we can boost the returns of our farm products and, therefore, the economy of our nations. Moreover, by establishing strong links between farmers and farmers, these systems make agriculture simple and inhibit the transformation of the agricultural land to residential areas [9]. The presence of oxygen and CO2 was measured by nano material film-based gas sensors. The presence of gases will be detected and the increase or decrease of graph-enriched polymer films interspersed in an embedded system, especially the gas level in the environment (**Figures 3** and **4**).

3. Interlighting odules

Power growers to produce exceptional returns of products with attractive characteristics, using industrial plants with sunlight-based light. In winter, when the sole-based elevation and the PPFD overhang are low and day long is smaller than in summer, a crop that is of high planting thickness, low photosynthetic moving thickness (PPFD) in the lower leaves generally is cut off in the plant development. This allows additional lighting to build profitability for the lower shelter throughout the year. Additional lighting can be expensive in any case. In some places the cost of power is lower around the evening, but there is no inspection of the effect of using additional light around the evening [21]. In this study, the effects of additional LEDs between lighting during the day and evening were examined for photosynthesis, development, and yields in both winter and summer between lighting modules with a rounded red and blue light to enlighten the lower leaves directly following the initial proposal [12]. LED was used between lighting modules with a consolidated red and blue light. The LED module was estimated at the PPFD of the light at 165 μmolm(−2)s(−1). Driven between the illumination, daytime between 4:00 am and 4:00 pm and night between 10:00 pm and 10:00 am.

Plants which were clearly represented by sunlight were used as controls. The photosynthetic limit of the centre and the lower leaves between lights expanded with daytime LEDs, which in winter total expanded yield by 27%; however, in the summer photosynthesis limit and yield were not expanded substantially. This methodology allows the yield to be increased by 24% in winter and 12% in summer. What's more, evening LED between lighting in winter essentially expanded the absolute solvent solids and ascorbic corrosive substance of the tomato organic products, by 20 and 25%, separately. The use of LEDs between lighting during the evening was also financially more knowledgeable than between lighting at daytime. This allows LEDs between the lighting in the evening to improve the development of tomato plant and yield, as well as the daytime and summer costs.

The understory leaves were enlightened by the intermediate lights (Philips Green Power LED between DR/B, Philips, Eindhoven, the Netherlands). The light range was red and blue together with a PPFD measured at 10 cm from the LED module, of 165 μmol m-2 s-1, and 40 cm from the Styrofoam board under which the root frame has been built.

3.1 Distribution of light in the plant profile

At each level of cover (top, centre, and bottom) a quantum sensor estimated the light transmission along the plant profile (LI-190SA; Li-Cor). The sensor was located with the ultimate goal to equivalent the trend edge to the leaves of the agent protection. LEDs between light estimates were made while LEDs were used between lighting. As control, sun-powered irradiance alone was used and without an LED between lights was estimated.

In winter, light is not standing and influences photosynthesis and production in those lines, as the development of plants and yield generally relies on photosynthesis (Hao and Papadopoulos, 1999). (Yamori, 2013; Yamori and Shikanai, 2016; Yamori et al., 2016). Since the development of one-bracket uses high plant thickness, light for the lower covering blocks is a significant restrictive variable (Lu et al., 2012a). In winter, an attempt at light capture by sun-based light is restricted in both the top and the base shelter leaves, whilst in the summer; the lower overhang leaves are constrained by light due to the thickness of the high plants (Gunnlaugsson and Adalsteinsson, 2006). (Ackerly and Bazzaz, 1995). In all, a 1% decrease in the total daily light throughout the developing season results in a loss of 1% in childcare production (Cockshull et al., 1992).

High planting thickness reduces light spread and the plant profile connected with common concealment (Zhang et al., 2015). The understories of tomatoes are extremely low net photosynthetic rate, due to their lesser light and leaf senescence (Acock et al., 1978; Xu et al., 1997) However, the low transmittal light that results in senescence affected the photosynthetic pace of the understory leaves (Acock et al., 1978). Frantz et al. (2000) suggested the essentially postponement of further light inside cowpea, which cover the inner leaves. Increased lighting from below also prevented external leaves from spreading and expanded the photosynthesis rate, which improved the entire cultivation of leettuce (Zhang et al., 2015).

Between the light, additional high installed lighting can be more attractive (Adams et al. 2002), improving the net photo synthesis of the undercover and then the output (Hovi et al., 2004; Pettersen et al., 2010). More generally, half improves have been achieved in various crops, although certain tests on different plants and areas have demonstrated no increase in yield (Hovi et al. 2004; Hovi and Tahvonen, 2008; Pettersen et al. 2010; Lu et al. 2012a, b) (Gunnlaugsson and Adalsteinsson, 2006; Heuvelink et al., 2006; Trouwborst et al., 2010). In fact, the development of single-substantial tomatoes has increased by 20% in winter between lighting output and by 14% in harvest time. (Second Words, 2012a,b). Some cover can then be useful to illuminate the lower section.

LEDs are seen as a suitable light hotspot between lighting (Hao et al. 2012) on the ground that they produce less warmth and are less likely to consume leaves as contrasting and HPS lights. In the previous decades, the progression of LEDs as an elective lighting source has enabled scientists and rancher to control their phantom characteristics through the consolidation of different light sources with various discharges on wavebands (Goto 2003; Merrill and al. 2016). There has been discussion of efficient plant development phases and light properties between the lighting application (Lu et al., 2012a, b); however, no research has been made into improvements in the lighting time frame with additional lights between winter and summer.

Moreover, the vitality production use of lighting can be accomplished by changing the LEDs between the illumination plans to use it further in the nighttime since the costs per unit kilowatt can be reduced with the off-the-run time of use (TOU). Different companies have recharged their electricity expenses with limited off-top rates (Ashok and Banerjee, 2000; Ashok, 2006; Middelberg et al., 2009). In the creation of nursery crops, farmers are concerned with elective lighting systems, which can produce yield and reduce labour costs. In all events, no analysis has been taken into account of the effect of supplemental evening time between photosynthesis, development and output lighting. In this study, we analysed the impact of LEDs during the day or evening between photosynthesis lighting, the development and yield in winter and summer in a single bracket of tomatoes.

Connected with the red joined blue light between lighting modules to illuminate lower leaves. Graph / Sun-oriented cells could improve this.

Graph-based sunlight-based sunlight cells could be used to improve the skills of sun-based cells and to gain vitality from these sun-based cells. New graph/si Schottky intersectional sun - powered cell with a backrest contact structure that has the advantages of easier production, lower creative costs, and greater dynamic area when examined using a gadget produced with the previous structure. Such sun-powered cell varieties had been delivered and therefore the efficiency is increased.

Sun powered boards work best today in times of strong daylight, but start dying when it gets cloudy or blistering. A forward step in sun powered boards based on Graphen could make a difference by allowing sun-based boards to create power in the disastrous environment. Finally, profitability, vitality collection and water use were retouched by the sending of information about the GSM condition without human intercession. Efficacy will also enhance. Great acres of land can be maintained with less human maintenance and also an enormous increase in efficiency.

4. Energy harvesting module

Sunlight based vitality is broadly accessible vitality source on the planet. Sun - based force isn't just acceptable by the perspective on the economy yet additionally it is condition well disposed type of the vitality. Presently days this vitality is utilized in road light and in other local burdens. In the present life because of advance innovation's the expense of sunlight based board diminishes, that will assist with utilizing sun based vitality in different segments. One of the uses of sun - powered vitality is in the water system framework. In India there is serious issue of vitality, in this manner sun - powered vitality is best answer for Indian rancher. The consistent extraction of water from earth is resultant into decline in water level from the earth with the goal that part of land comes gradually in the un-flooded zone, another explanation of this is because of impromptu water system. Additionally now-a-day's populace increments quickly so request of food likewise builds which doesn't get balance among request and flexibly of food.

4.1 Concept of smart irrigation

The old water system techniques are sprinklers and flood type framework. In these strategies, the utilization of water is in enormous sum. On account of inclines in the field a huge measure of water moves downwards. In this manner, the rest of the piece of the field remains un-flooded. Huge measure of water goes squander in these strategies. Such an issue could be overwhelmed by this work which utilizes sensors with microcontroller, subsequently half water sparing is accomplished. Utilization of sun oriented board makes this green method of vitality sparing [22].

As per the study directed by the Bureau of Electrical Energy, in India in 2011 there are around 18 million farming siphon sets and around 0.5 million new associations for every Sardesai Mayur A. Patil Ranjeet G. Patil Ranjit B. Katkar Kiran B. Sutar Rohit R. Dr. IrranaKorachgoan Department of Electrical Engineering, Shivaji University/AMGOI Wathar, Kolhapur, Maharashtra, India,are introduced with normal limit of 5HP. Complete yearly utilization in the horticulture division is 131.96 billion KWh. (19% of all out power utilization) [17].

METHODOLOGY: Proposed water system framework comprises of two sections, sunlight - based siphoning and programmed water system part. Sun based board charges the battery through charge controller. From the battery, flexibly is given to the engine legitimately in this work [17].

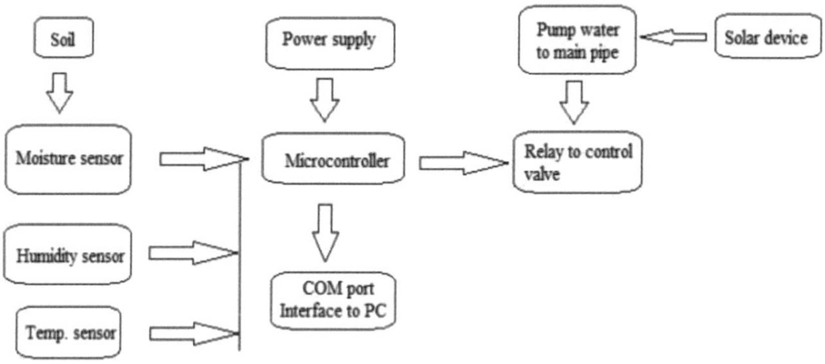

Figure 5.
Block diagram of solar powered irrigation system.

Figure 5 shows the square graph demonstrating the fundamental pieces of sun based fueled water system framework. Here the detecting circuit controls the engine. The sensors utilized are soil dampness sensor, temperature and mugginess sensor. The sensor distinguishes the estimations of soil dampness, temperature and stickiness at various focuses in the field. Microcontroller as indicated by pre-set worth looks at the deliberate qualities. In light of the blunder between the pre-set and estimated values, engine ON/OFF condition is controlled [23].

4.2 Solar irrigation system

The siphons utilized for the vehicle of the water are outfitted with sunlight based cells. The sun- powered vitality consumed by the cells is then changed over into electrical vitality through a generator which at that point takes care of an electric engine driving the siphon. The vast majority of the conventional siphon frameworks for the most part work with a diesel motor or with the nearby force network. In any case, these two methods of tasks present inconveniences contrasted with sunlight based siphons [24].

In numerous country territories, particularly in creating and developing nations, the entrance to the power network isn't constantly ensured. For this situation, ranchers can't depend on the conventional water system framework. Hence, utilizing an autonomous and elective vitality framework can be an answer for the rancher to make sure about a protected force source and for the open lattice to maintain a strategic distance from immersion.

Diesel siphons are somewhat more effective than AC fueled siphons as they permit more prominent adaptability. In any case, one of the fundamental imperatives is that this framework depends on the fuel accessibility, included to a more noteworthy effect the earth. Diesel-driven siphons are less expensive than sun oriented controlled siphons yet the working expenses are very high and rely vigorously upon the diesel cost. In sunlight based controlled frameworks, it works the other path round, that is, in spite of the fact that this framework is moderately costly, the wellspring of vitality is free, accordingly, after the amortization time frame, there are done working costs (just the upkeep costs must be thought of). Accordingly, sun oriented siphons end up being a feasible long haul venture.

As a few investigations, such as Water for riches and food security by AgWater Solutions Project, have appeared, the entrance to water for agrarian purposes stays basic in certain zones, for example, in dry districts of Africa and Southern Asia. Numerous Indian and African ranchers bring the water straightforwardly from the well or the waterways and inundate their fields utilizing cans. In the event that ranchers of those areas could approach a mechanized siphon, they would increase their yield by 300%.

As a result, R&D is nowadays generally focusing on sun-based syphons which in parched locations are moderate. In collaboration with Siemens, the company IBC SOLAR has created a response to the sun-based diesel engine replacement. The entire water system framework, including the syphon is worth in this situation; only a photovoltaic frame and the so-called IBC syphon control system replace the diesel engine. In 2015, a model was tested and was ultimately extremely effective for this framework from a ranch in Namibia, as the producers had pointed out. The main leeway lies in the way that the current framework utilises low costs of obtaining. The controller consists of 10 simple information channels, 6 simple delivery channels, 40 I/O lines, 3 accelerometer implants and WiFi. The advanced and simple data sources are used for the interface with the controls, the soil moisture sensor, the adhesive and temperature sensors and the current rates sensors. The transmission panel is used to interface the syphon and a saphon that controls the process of the water system. The transmission panel operates at 5 V flexible voltage. The 5 V sign is supplied by a buck converter to the transmission board, which sets the 12 V of the load controller down to the ideal 5 V voltage.

A model has been structured and tested. The upper part is a rectangular tub with double layer loaded with soil, while the lower part is loaded with water to imitate the underground water table completely or uncompletely. The plastic layer, which isolates the top and ground layers, has openings which allow water to flow into the bottom layer which has not been consumed by the dirt. A syphon is lowered into a subterranean water table that concentrates and stores water in the water supply (water stockpiling tank). The bilge syphon channel has a channel that is not sucked into the syphon by broken soil or by any other particulary material. This general structure is designed to emulate ranches which can approach ground water but can access almost zero water sources. In the process of the dribble water system described in **Figure** 7, a stomach syphon separates the water from the tank and is responsible for flooding the household. The water rate can be controlled through the stomach syphon, which controls the water rate by irrigation dribble [25].

4.3 Renewable energy requirement calculation

This segment presents the computations required to evaluate the force expected to work the proposed keen water system framework. As referenced before, for this model ranch plan, it is accepted that the homestead will have 5.5 h of steady splendid daylight daily, i.e. HSD = 5.5 h.

It is additionally accepted that solitary 80% of this 5.5-h range produces useable force, i.e. ηusable = 0.8. Note that the numbers for HSD and ηusable are not picked totally discretionarily. This can be seen by playing out the accompanying calculation. Think about that as a day has 12 h of daylight (paying little heed to splendor), and afterward figure the proportion $(5.5 \times 0.8)/12 = 36.67\%$.

This shows regardless of whether a sun based board is set out in the sun for the whole day, just about 37% of the hours of the whole day add to useable electrical vitality, which might be a traditionalist figure giving the condition of at present proceeding with mechanical headways identified with sunlight based boards, and thus the above suppositions appear to be sensible.

The accompanying figuring's give subtleties identified with the choice of the battery required for the proposed model of the brilliant water system framework. For a 100 W sun based board, as this rating is bigger than the determined98.6 W rating. It is worth to take note of that the force determined is the force required for the one-day activity, and not prompt force required.

The everyday vitality required to be provided/put away by the battery Edbatt can be figured as the proportion the vitality required every day separated by the battery proficiency,

Edbatt = Edtot/ηbatt = 369/0.8 = 461.25 Wh/d.

The useable Ampere-hour (Ah) limit of the necessary battery can be determined by partitioning Edbatt by the evaluated terminal voltage of the battery, for example.

Cbattday = Edbatt/Vbatt = 461.25/12 = 38:44 Ah.

The usable limit required for the battery, to keep up the activity for a day is, 38.44/0.75 = 51.26 Ah.

The battery should not be completely released as its life span will be shortened; the depth of release will be considered afterwards. If the sonny-light board is short or does not reach daylight for the whole day, the battery is fully charged on the previous day by using a battery with this limit. As mentioned above, a 100 W PV sun-based panel and a deep cycle, lead-corrosive battery with limit of 55 A h, evaluated at 12 V, is to be used as mentioned previously. In view of the fact that its structure makes it possible to charge and release different occasions without having a complete impact on the battery health a profound cycle battery is used [24].

5. Data acquisition, processing and recording

5.1 Simulation results and validation

The circuit is recreated using the PROTEUS system for electronic testing. This model produces the ground-humidity sensor, the light sensor and the temperature sensor as microcontroller feeds and comparative results on the LCD. **Figure 6** shows the sensor of soil humidity detection. In this structure, the ideal soil moisture level is set to 100 volts. The process is started and the motor activated if the voltage level is not exactly the water system limit (100 Volts). When the level exceeds the rim, the engine stops and the control is naturally resolved without human mediation. The engine is not controlled [9].

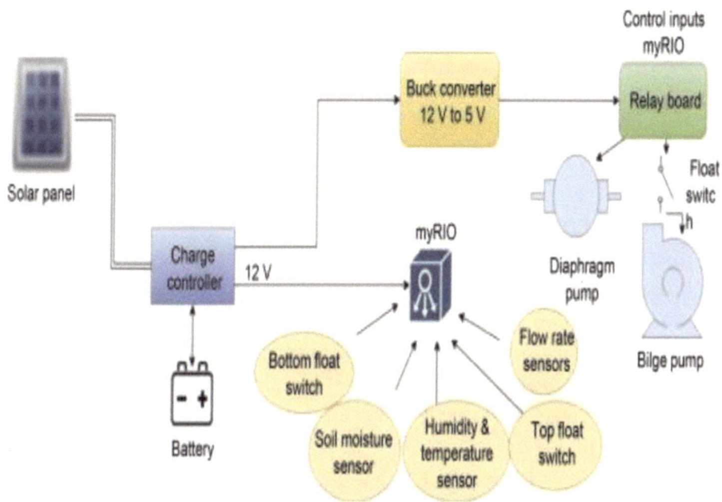

Figure 6.
Overall system design.

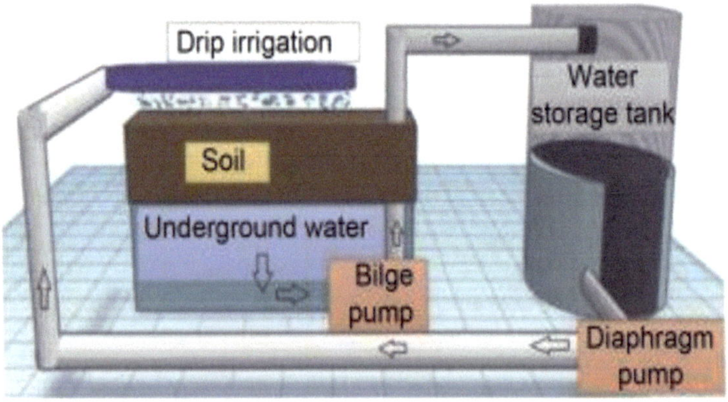

Figure 7.
Physical layout of the prototype farm and smart irrigation system.

The controller consists of 10 simple information channels, 6 simple delivery channels, 40 I/O lines, 3 accelerometer implants and WiFi. The advanced and simple data sources are used for the interface with the controls, the soil moisture sensor, the adhesive and temperature sensors and the current rates sensors. The transmission panel is used to interface the syphon and a saphon that controls the process of the water system. The transmission panel operates at 5 V flexible voltage. A buck converter is given to this 5 V sign to the Transfer Board, which descends the 12 V voltage flexibly from the load controller to the ideal 5 V level.

A model has been structured and tested. The upper part is a rectangular tub with double layer loaded with soil, while the lower part is loaded with water to imitate the underground water table completely or uncompletely. The plastic layer, which isolates the top and ground layers, has openings which allow water to flow into the bottom layer which has not been consumed by the dirt. A syphon is lowered into a subterranean water table that concentrates and stores water in the water supply (water stockpiling tank). The bilge syphon channel has a channel that is not sucked into the syphon by broken soil or by any other particulary material. This general structure is designed to emulate ranches which can approach ground water but can access almost zero water sources. In the process of the dribble water system described in **Figure 7**, a stomach syphon separates the water from the tank and is responsible for flooding the household. The water rate can be controlled through the stomach syphon, which controls the water rate by irrigation dribble.

5.2 Renewable energy requirement calculation

This segment presents the computations required to evaluate the force expected to work the proposed keen water system framework. As referenced before, for this model ranch plan, it is accepted that the homestead will have 5.5 h of steady splendid daylight daily, i.e. HSD = 5.5 h.

It is additionally accepted that solitary 80% of this 5.5-h range produces useable force, i.e. ηusable = 0.8. Note that the numbers for HSD and ηusable are not picked totally discretionarily. This can be seen by playing out the accompanying calculation. Think about that as a day has 12 h of daylight (paying little heed to splendor), and afterward figure the proportion $(5.5 \times 0.8)/12 = 36.67\%$.

This shows regardless of whether a sun based board is set out in the sun for the whole day, just about 37% of the hours of the whole day add to useable electrical vitality, which might be a traditionalist figure giving the condition of at present proceeding with mechanical headways identified with sunlight based boards, and thus the above suppositions appear to be sensible.

The accompanying figuring's give subtleties identified with the choice of the battery required for the proposed model of the brilliant water system framework. For a 100 W sun based board, as this rating is bigger than the determined 98.6 W rating. It is worth to take note of that the force determined is the force required for the one-day activity, and not prompt force required.

The everyday vitality required to be provided/put away by the battery Edbatt can be figured as the proportion of the vitality required every day separated by the battery proficiency,

Edbatt = Edtot/ηbatt = 369/0.8 = 461.25 Wh/d.

The useable Ampere-hour (Ah) limit of the necessary battery can be determined by partitioning Edbatt by the evaluated terminal voltage of the battery, for example.

Cbattday = Edbatt/Vbatt = 461.25/12 = 38:44 Ah.

The usable limit required for the battery, to keep up the activity for a day is, 38.44/0.75 = 51.26 Ah.

The battery ought not to be released totally since it will abbreviate its life expectancy; subsequently, the insight of release is thought about [23, 24]. Utilizing a battery of this limit resolve the framework for a whole day if the sunlight based board comes up short or doesn't get daylight for the entire day, accepting the battery was charged to the full limit on the previous day [15]. As indicated by the above estimations, it is chosen to utilize a 100 W PV sun based board as referenced previously, and a profound cycle, lead-corrosive battery of the limit of 55 A h evaluated at 12 V. A profound cycle battery is utilized, in light of the fact that its structure empowers it to be charged and released various occasions without altogether influencing the general battery health [21].

Author details

Arasu Sivagami[1*], Michael Angelo Kandavalli[2] and Bhaskarrao Yakkala[1]

1 Department of NEMS, Institute of ECE, Saveetha School of Engineering, SIMATS, Chennai, Tamilnadu, India

2 Department of ECE, Dhanekula Institute of Engineering and Technology, Vijayawada, Andhra Pradesh, India

*Address all correspondence to: sivagamiarasu.sse@saveetha.com

References

[1] E.Sowmiya, S.Sivaranjani Smart System Monitoring on Soil Using Internet of Things (IOT) International Research Journal of Engineering and Technology (IRJET), Volume: 04 Issue: 02 Feb (2017)

[2] A.Sivagami,U.Hareeshvare, S.Maheshwar,Dr.V.S.K.Venkatachalapathy, "Automated Irrigation System for Greenhouse Monitoring", in Journal. 2 of Instituition of Engineers (India): Series A, https://doi.org/10.1007/s40030-018-0264-0018)

[3] Ch Sumaliya, C Bharatender Rao "Smart Farm Monitoring Using Raspberry Pi and Arduino", International Journal of Management Studies (IJMS), Volume: 01 Issue: 11 | Nov (2016)

[4] M. Srbinovska, C. Gavrovski, V. Dimcev, A. Krkoleva, V. Borozan, "Environmental parameters monitoring in precision agriculture using wireless sensor networks" J. Clean. Prod., 88(2015), pp. 297-307

[5] www.youtube.com

[6] S. Darshna, T.Sangavi,Sheena Mohan, A.Soundharya, Sukanya Desikan "Smart irrigationSystem" IOSR Journal of Electronics and Communication Engineering (IOSR-JECE) , Volume 10, Issue 3, Ver. II May -Jun.2015

[7] Sneha Angal, "Raspberry pi and Arduino Based Automated Irrigation System" Department of Electronics & telecommunication, Dhole Patil College of Engineering, Pune, India

[8] Sudhir Rao Rupanagudi, Ranjani B. S., Prathik Nagaraj, Varsha G Bhat, and Thippeswamy G, "A Novel Cloud Computing based Smart Farming System for Early Detection of Borer Insects in Tomatoes," ICCICT, pp.1-6, 2015

[9] Zulkifli, C. Z.* and Noor, N. N. "Wireless Sensor Network and Internet of Things (IoT) Solution in Agriculture" Pertanika J. Sci. & Technol. 25 (1): 91 - 100 (2017)

[10] Prabhu, Boselin and Pradeep, M. and Gajendran, E., "An Analysis of Smart Irrigation System Using Wireless Sensor Network" Star Vol.5 Issue 3(3), March (2017)

[11] Chandan Kumar, pramiteebehera "A Low Cost Smart Irrigation Control System", International Conference on Electronics and Communication System (ICECS 2015) IEEE 1146

[12] Pranita A. Bhosale, Prof. V. V. Dixit, Water Saving-Irrigation Automatic Agricultural Controller, International Journal of Scientific & Technology Research volume 1, Issue 11,December 2012

[13] Lala Bhaskar, BarkhaKoli, Punit Kumar, Vivek Gaur, "Automatic Crop Irrigation System" IEEE (2015).

[14] Ravi Kishore Kodali, Borade Samar Sarjerao "A Low Cost Smart Irrigation System using MOTT Protocol", IEEE 2017

[15] J.J. Rodriguez-Andina, M.J. Moure, M.D. ValdesFeatures, design tools, and application domains of FPGAs IEEE Trans. Ind. Electron., 54 (4) (Aug. 2007), pp. 1810-1823

[16] https://www.google.com

[17] S. Harishankar, R. Sathish Kumar, SudharsanK.P, U. Vignesh and T.Viveknath,2014, Solar Powered Smart Irrigation System, Advance in Electronic and Electric Engineering. ISSN 2231-1297, Volume 4, Number 4, pp. 341-346, Research India Publications.

[18] Md Saifudaullah Bin Baharudin and Rosnin Abu Kassim, "Development of

Fire Alarm System using Raspberry Pi and Arduino Uno" 2013 International Conference on Electrical,Electronics and System Engineering.

[19] Jainishkuamr Anghan1, Parveen Sultana H1,Smart Irrigation System using Raspberry Pi,International Journal of Scientific & Engineering Research, Volume 9, Issue 6, June-2018,ISSN 2229-5518.

[20] PothabathinaShirisha, Jammula Madhuri, Irrigation System Using Raspberry PI and GSM,IJMETR Journal, ISSN 2348-4845,pp.54-61.

[21] J.J. Rodríguez-Andina, M.D. Valdés-Peña, M.J. Moure,Advanced features and industrial applications ofFPGAs—a review, IEEE Trans. Ind. Inf., 11 (4) (Aug. 2015), pp. 853-864.

[22] K. Prathyusha and Chaitanya Suman, 2012, Design of embedded systems for the automation of drip irrigation, International Journal of Application or Innovation in Engineering & Management (IJAIEM), Volume 1, Issue 2. pp. 254-258.

[23] Satyendra Tripathi, Lakshmi N., Sai Apoorva and U. A. Vasan, Solar powered intelligent drip irrigation system for sustainable irrigation services, pp-1-8.

[24] A.R.Al Ali etal,IoT-solar energy powered smart farm irrigation system,Journal of Electronic Science and Technology, Volume 17, Issue 4, December 2019

[25] Ryu M, Yun J, Miao T, Ahn IY, Choi SC, Kim J (2015) "Design and implementation of a connected farm for smart farming system" 2015 IEEE, pp 1-4 (2015).